Teacher, Student, and Parent
One-Stop Internet Resources

Log on to
booke.msscience.com

ONLINE STUDY TOOLS

- Section Self-Check Quizzes
- Interactive Tutor
- Chapter Review Tests
- Standardized Test Practice
- Vocabulary PuzzleMaker

ONLINE RESEARCH

- WebQuest Projects
- Prescreened Web Links
- Career Links
- Internet Labs

INTERACTIVE ONLINE STUDENT EDITION

- Complete Interactive Student Edition available at mhln.com

FOR TEACHERS

- Teacher Bulletin Board
- Teaching Today—Professional Development

SAFETY SYMBOLS

SAFETY SYMBOLS	HAZARD	EXAMPLES	PRECAUTION	REMEDY
DISPOSAL	Special disposal procedures need to be followed.	certain chemicals, living organisms	Do not dispose of these materials in the sink or trash can.	Dispose of wastes as directed by your teacher.
BIOLOGICAL	Organisms or other biological materials that might be harmful to humans	bacteria, fungi, blood, unpreserved tissues, plant materials	Avoid skin contact with these materials. Wear mask or gloves.	Notify your teacher if you suspect contact with material. Wash hands thoroughly.
EXTREME TEMPERATURE	Objects that can burn skin by being too cold or too hot	boiling liquids, hot plates, dry ice, liquid nitrogen	Use proper protection when handling.	Go to your teacher for first aid.
SHARP OBJECT	Use of tools or glassware that can easily puncture or slice skin	razor blades, pins, scalpels, pointed tools, dissecting probes, broken glass	Practice common-sense behavior and follow guidelines for use of the tool.	Go to your teacher for first aid.
FUME	Possible danger to respiratory tract from fumes	ammonia, acetone, nail polish remover, heated sulfur, moth balls	Make sure there is good ventilation. Never smell fumes directly. Wear a mask.	Leave foul area and notify your teacher immediately.
ELECTRICAL	Possible danger from electrical shock or burn	improper grounding, liquid spills, short circuits, exposed wires	Double-check setup with teacher. Check condition of wires and apparatus.	Do not attempt to fix electrical problems. Notify your teacher immediately.
IRRITANT	Substances that can irritate the skin or mucous membranes of the respiratory tract	pollen, moth balls, steel wool, fiberglass, potassium permanganate	Wear dust mask and gloves. Practice extra care when handling these materials.	Go to your teacher for first aid.
CHEMICAL	Chemicals can react with and destroy tissue and other materials	bleaches such as hydrogen peroxide; acids such as sulfuric acid, hydrochloric acid; bases such as ammonia, sodium hydroxide	Wear goggles, gloves, and an apron.	Immediately flush the affected area with water and notify your teacher.
TOXIC	Substance may be poisonous if touched, inhaled, or swallowed.	mercury, many metal compounds, iodine, poinsettia plant parts	Follow your teacher's instructions.	Always wash hands thoroughly after use. Go to your teacher for first aid.
FLAMMABLE	Flammable chemicals may be ignited by open flame, spark, or exposed heat.	alcohol, kerosene, potassium permanganate	Avoid open flames and heat when using flammable chemicals.	Notify your teacher immediately. Use fire safety equipment if applicable.
OPEN FLAME	Open flame in use, may cause fire.	hair, clothing, paper, synthetic materials	Tie back hair and loose clothing. Follow teacher's instruction on lighting and extinguishing flames.	Notify your teacher immediately. Use fire safety equipment if applicable.

 Eye Safety Proper eye protection should be worn at all times by anyone performing or observing science activities.

 Clothing Protection This symbol appears when substances could stain or burn clothing.

 Animal Safety This symbol appears when safety of animals and students must be ensured.

 Handwashing After the lab, wash hands with soap and water before removing goggles.

Glencoe Science

Ecology

NATIONAL GEOGRAPHIC

Glencoe

New York, New York   Columbus, Ohio   Chicago, Illinois   Woodland Hills, California

Ecology

Three generations of reticulated giraffes are interacting on an African savanna. Females start breeding at around five years old, and gestation is about 15 months. Calves are born from a standing female, dropping six feet to the ground! They weigh 100 to 150 pounds, and are six feet tall at birth.

 **Glencoe**

The *McGraw·Hill* Companies

Send all inquiries to:
Glencoe/McGraw-Hill
8787 Orion Place
Columbus, OH 43240-4027

ISBN: 978-0-07877820-9
MHID: 0-07-877820-4

Printed in the United States of America.

4 5 6 7 8 9 10 DOW 09

Authors

NATIONAL GEOGRAPHIC
Education Division
Washington, D.C.

Peter Rillero, PhD
Associate Professor of
Science Education
Arizona State University West
Phoenix, AZ

Dinah Zike
Educational Consultant
Dinah-Might Activities, Inc.
San Antonio, TX

Series Consultants

CONTENT

Michael A. Hoggarth, PhD
Department of Life and Earth Sciences
Otterbein College
Westerville, OH

Dominic Salinas, PhD
Middle School Science Supervisor
Caddo Parish Schools
Shreveport, LA

MATH

Teri Willard, EdD
Mathematics Curriculum Writer
Belgrade, MT

READING

Elizabeth Babich
Special Education Teacher
Mashpee Public Schools
Mashpee, MA

SAFETY

Sandra West, PhD
Department of Biology
Texas State University-San Marcos
San Marcos, TX

ACTIVITY TESTERS

Nerma Coats Henderson
Pickerington Lakeview Jr. High School
Pickerington, OH

Mary Helen Mariscal-Cholka
William D. Slider Middle School
El Paso, TX

Science Kit and Boreal Laboratories
Tonawanda, NY

Series Reviewers

Maureen Barrett
Thomas E. Harrington Middle School
Mt. Laurel, NJ

Desiree Bishop
Environmental Studies Center
Mobile County Public Schools
Mobile, AL

Linda V. Forsyth
Retired Teacher
Merrill Middle School
Denver, CO

Amy Morgan
Berry Middle School
Hoover, AL

Darcy Vetro-Ravndal
Hillsborough High School
Tampa, FL

HOW TO...
Use Your Science Book

Before You Read

- **Chapter Opener** Science is occurring all around you, and the opening photo of each chapter will preview the science you will be learning about. The **Chapter Preview** will give you an idea of what you will be learning about, and you can try the **Launch Lab** to help get your brain headed in the right direction. The **Foldables** exercise is a fun way to keep you organized.

- **Section Opener** Chapters are divided into two to four sections. The **As You Read** in the margin of the first page of each section will let you know what is most important in the section. It is divided into four parts. **What You'll Learn** will tell you the major topics you will be covering. **Why It's Important** will remind you why you are studying this in the first place! The **Review Vocabulary** word is a word you already know, either from your science studies or your prior knowledge. The **New Vocabulary** words are words that you need to learn to understand this section. These words will be in **boldfaced** print and highlighted in the section. Make a note to yourself to recognize these words as you are reading the section.

Ecology

As You Read

- **Headings** Each section has a title in large red letters, and is further divided into blue titles and small red titles at the beginnings of some paragraphs. To help you study, make an outline of the headings and subheadings.

- **Margins** In the margins of your text, you will find many helpful resources. The **Science Online** exercises and **Integrate** activities help you explore the topics you are studying. **MiniLabs** reinforce the science concepts you have learned.

- **Building Skills** You also will find an **Applying Math** or **Applying Science** activity in each chapter. This gives you extra practice using your new knowledge, and helps prepare you for standardized tests.

- **Student Resources** At the end of the book you will find **Student Resources** to help you throughout your studies. These include **Science, Technology,** and **Math Skill Handbooks,** an **English/Spanish Glossary,** and an **Index.** Also, use your **Foldables** as a resource. It will help you organize information, and review before a test.

- **In Class** Remember, you can always ask your teacher to explain anything you don't understand.

FOLDABLES™
Study Organizer

Science Vocabulary Make the following Foldable to help you understand the vocabulary terms in this chapter.

STEP 1 Fold a vertical sheet of notebook paper from side to side.

STEP 2 Cut along every third line of only the top layer to form tabs.

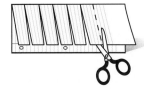

STEP 3 Label each tab with a vocabulary word from the chapter.

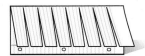

Build Vocabulary As you read the chapter, list the vocabulary words on the tabs. As you learn the definitions, write them under the tab for each vocabulary word.

Look For...
FOLDABLES™
At the beginning of every section.

In Lab

Working in the laboratory is one of the best ways to understand the concepts you are studying. Your book will be your guide through your laboratory experiences, and help you begin to think like a scientist. In it, you not only will find the steps necessary to follow the investigations, but you also will find helpful tips to make the most of your time.

- Each lab provides you with a **Real-World Question** to remind you that science is something you use every day, not just in class. This may lead to many more questions about how things happen in your world.

- Remember, experiments do not always produce the result you expect. Scientists have made many discoveries based on investigations with unexpected results. You can try the experiment again to make sure your results were accurate, or perhaps form a new hypothesis to test.

- Keeping a **Science Journal** is how scientists keep accurate records of observations and data. In your journal, you also can write any questions that may arise during your investigation. This is a great method of reminding yourself to find the answers later.

Look For...

- **Launch Labs** start every chapter.
- **MiniLabs** in the margin of each chapter.
- **Two Full-Period Labs** in every chapter.
- **EXTRA Try at Home Labs** at the end of your book.
- the **Web site** with laboratory demonstrations.

Before a Test

Admit it! You don't like to take tests! However, there *are* ways to review that make them less painful. Your book will help you be more successful taking tests if you use the resources provided to you.

- Review all of the **New Vocabulary** words and be sure you understand their definitions.

- Review the notes you've taken on your **Foldables,** in class, and in lab. Write down any question that you still need answered.

- Review the **Summaries** and **Self Check questions** at the end of each section.

- Study the concepts presented in the chapter by reading the **Study Guide** and answering the questions in the **Chapter Review.**

Look For...

- **Reading Checks** and **caption questions** throughout the text.
- the **Summaries** and **Self Check questions** at the end of each section.
- the **Study Guide** and **Review** at the end of each chapter.
- the **Standardized Test Practice** after each chapter.

Let's Get Started

To help you find the information you need quickly, use the Scavenger Hunt below to learn where things are located in Chapter 1.

1. What is the title of this chapter?

2. What will you learn in Section 1?

3. Sometimes you may ask, "Why am I learning this?" State a reason why the concepts from Section 2 are important.

4. What is the main topic presented in Section 2?

5. How many reading checks are in Section 1?

6. What is the Web address where you can find extra information?

7. What is the main heading above the sixth paragraph in Section 2?

8. There is an integration with another subject mentioned in one of the margins of the chapter. What subject is it?

9. List the new vocabulary words presented in Section 2.

10. List the safety symbols presented in the first Lab.

11. Where would you find a Self Check to be sure you understand the section?

12. Suppose you're doing the Self Check and you have a question about concept mapping. Where could you find help?

13. On what pages are the Chapter Study Guide and Chapter Review?

14. Look in the Table of Contents to find out on which page Section 2 of the chapter begins.

15. You complete the Chapter Review to study for your chapter test. Where could you find another quiz for more practice?

Teacher Advisory Board

The Teacher Advisory Board gave the editorial staff and design team feedback on the content and design of the Student Edition. They provided valuable input in the development of the 2008 edition of *Glencoe Science.*

John Gonzales
Challenger Middle School
Tucson, AZ

Rachel Shively
Aptakisic Jr. High School
Buffalo Grove, IL

Roger Pratt
Manistique High School
Manistique, MI

Kirtina Hile
Northmor Jr. High/High School
Galion, OH

Marie Renner
Diley Middle School
Pickerington, OH

Nelson Farrier
Hamlin Middle School
Springfield, OR

Jeff Remington
Palmyra Middle School
Palmyra, PA

Erin Peters
Williamsburg Middle School
Arlington, VA

Rubidel Peoples
Meacham Middle School
Fort Worth, TX

Kristi Ramsey
Navasota Jr. High School
Navasota, TX

Student Advisory Board

The Student Advisory Board gave the editorial staff and design team feedback on the design of the Student Edition. We thank these students for their hard work and creative suggestions in making the 2008 edition of *Glencoe Science* student friendly.

Jack Andrews
Reynoldsburg Jr. High School
Reynoldsburg, OH

Peter Arnold
Hastings Middle School
Upper Arlington, OH

Emily Barbe
Perry Middle School
Worthington, OH

Kirsty Bateman
Hilliard Heritage Middle School
Hilliard, OH

Andre Brown
Spanish Emersion Academy
Columbus, OH

Chris Dundon
Heritage Middle School
Westerville, OH

Ryan Manafee
Monroe Middle School
Columbus, OH

Addison Owen
Davis Middle School
Dublin, OH

Teriana Patrick
Eastmoor Middle School
Columbus, OH

Ashley Ruz
Karrer Middle School
Dublin, OH

The Glencoe middle school science Student Advisory Board taking a timeout at COSI, a science museum in Columbus, Ohio.

Contents

In each chapter, look for these opportunities for review and assessment:
- Reading Checks
- Caption Questions
- Section Review
- Chapter Study Guide
- Chapter Review
- Standardized Test Practice
- Online practice at **booke.msscience.com**

Get Ready to Read Strategies

Student Resources

Cross-Curricular Readings/Labs

Content Details

Labs/Activities

INTEGRATE

Science Online

Standardized Test Practice

Content Details

Conservation and Native Americans

Figure 1 Salmon were the major food source of the Pacific Northwest Native Americans.

A cool breeze blows through the trees lining the Columbia River in the state of Washington as shiny salmon dart through the water. Every year these fish make the difficult journey upstream to lay their eggs in the same waters where they began their lives.

Many years ago, before Europeans came to North America, Native Americans of the northwest depended on salmon. At the first sign of autumn, the salmon run would start. Native American oral histories tell of rivers being so full of the fish that you almost could walk across them to the other side. The native people harvested and dried enough fish to sustain them for the entire year. They also had a deep respect for the salmon. The belief that the fish had a spirit was shown by the tradition of thanking the salmon's spirit for its sacrifice before it was eaten.

Respect for Life

The Native American tradition is to respect the bodies and spirits of all animals that gave their lives to feed and clothe the people. On the Great Plains, the Cheyenne and other native peoples had strict rules against killing more bison than they needed. They believed in using every part of the animal. They used as much of the animal as possible for food. The animal's fat was used in cooking. Tools were made from bones. Clothing, shoes, and blankets were made from hides. Even the bisons' stomachs were used as water pouches.

Respect for life extended to plants and crops for Native Americans. The Iroquois of the northeastern United States celebrated festivals in honor of the "three sisters"—corn, squash,

Figure 2 Native Americans did not waste any part of nature. This Blackfoot shirt was made with hair, porcupine quills, and feathers.

and beans—their essential foods. The Maya who once lived in what is now southern Mexico and Central America, felt that if someone cut down a tree unnecessarily, that person's life would be shortened.

Every year, in the late spring or early summer, native peoples living on the Plains, including the Cree, Kiowa, Shoshone, and others, celebrated the cycle of life with the Sun Dance. This ceremony, which is performed by many tribes today, stresses the regeneration of life and humans' connection to Earth. Native Americans expressed thankfulness for Earth's gifts—the return of flowers and crops in the spring and summer, and the sacrifice made by animal spirits as they give their bodies to sustain the people. Their traditions recognize that people must cooperate with nature so that revival and rebirth can continue.

Figure 3 Agriculture students experiment with the Native American practice of growing several types of plants in one area.

Respect for Earth

Native Americans felt a deep connection to Earth. While some European settlers viewed the New World as a wilderness ready to be tamed, the native peoples believed in living in harmony with Earth. They didn't understand the European desire to own and develop land. To Native Americans, everyone shared the land. You could no more own the land than you could own the air. The Lakota chief Black Elk once spoke of each human's responsibility to Earth in this way: "Every step that we take upon You should be done in a sacred manner; every step should be taken as a prayer."

Figure 4 Native American artist Helen Hardin titled this piece *Father Sky Embracing Mother Earth.*

Figure 5 Early industrial development led to environmental pollution.

Need for Conservation

When Europeans began to settle throughout America, it seemed that their way of life would completely replace Native American conservation practices. Many farming techniques used by the new settlers eroded fertile topsoil. Some of the wildlife, which had been treasured and respected by the native peoples, was driven to either extinction or near extinction.

New technology greatly affected the environment. The steam and diesel engines, commercial oil drilling, burning of coal, and industrial development led to waste and pollution.

In recent years, Americans have become more aware of the need for conservation. In 1970, the Environmental Protection Agency was created with the goal of safeguarding the environment. Paper, plastic, glass, and metal are widely recycled. Many items that were once considered waste are now being reused. Although pollution is still a problem, Americans are returning to practices that are more like traditional Native American ways.

Science

The practice of modifying human behavior to preserve Earth is known as conservation. Conservation involves managing the use of natural resources found in many different environments. In this book, you will learn about how all species on Earth are connected and how they can affect their environments.

History of Science

History is a record of people's achievements and their mistakes. Science history teaches that scientific ideas are not limited to scientists. Many people of different ethnic backgrounds, professions, and ages—male and female—have contributed to today's conservation movement. Studying the history of science reminds people that even though accepted views may change, humans constantly gain a better understanding of nature.

Figure 6 Recycling and using nonpolluting types of energy, such as solar energy, are two of the ways Americans are becoming more environmentally conscious.

Connections

Throughout their history, Native Americans have understood the importance of living in harmony with nature. They practiced conservation as a way of life, not just a passing fad. The people of the northwest understood that fishing for more salmon than they needed would endanger the next year's supply. Today, people are learning this same lesson as some traditional fishing grounds are closed because of overharvesting.

Recall the Mayan belief about how unnecessarily cutting down a tree shortened one's life. It is now known that plants provide oxygen and remove carbon dioxide from the air. Scientists warn that cutting down rain forest trees might lead to a buildup of carbon dioxide in the atmosphere. This could add to the trend in rising, global average temperatures known as global warming.

Native Americans on the Plains realized the importance of using every part of the bison, learning to live with their minimum needs, and limiting their use of natural resources. This model is being followed today, in some ways, through the effort to reduce, reuse, and recycle.

As new knowledge replaces old, it's tempting to think that only new ideas are worthwhile. Native American traditions of conservation show that this is not always true. These traditions once were disregarded but the value of their conservation is now known.

Figure 7 Many items that you use every day are made of recycled material. These packing beads are made from a wheat product.

Figure 8 The amount of land covered by rain forests is decreasing daily.

You Do It

Forests originally covered about 25 percent of Earth's land areas. Today, only about 13 percent of Earth's land areas are covered with forests. Research to find out the history of this change. Explain how following Native American conservation practices might have avoided the problems that face forests today.

Interactions of Life

The BIG Idea

Living organisms interact with their environment and with one another in many ways.

SECTION 1
Living Earth
Main Idea All living and nonliving things on Earth are organized into levels, such as communities and ecosystems.

SECTION 2
Populations
Main Idea A population's size is affected by many things, including competition.

SECTION 3
Interactions Within Communities
Main Idea Every organism has a role in its environment.

Are these birds in danger?

The birds are a help to the rhinoceros. They feed on ticks and other parasites plucked from the rhino's hide. When the birds sense danger, they fly off, giving the rhino an early warning. Earth's living organisms supply one another with food, shelter, and other requirements for life.

Science Journal Describe how a familiar bird, insect, or other animal depends on other organisms.

Start-Up Activities

How do lawn organisms survive?

You probably have taken thousands of footsteps on grassy lawns or playing fields. If you look closely at the grass, you'll see that each blade is attached to roots in the soil. How do grass plants obtain everything they need to live and grow? What other kinds of organisms live in the grass? The following lab will give you a chance to take a closer look at the life in a lawn.

1. Examine a section of sod from a lawn.

2. How do the roots of the grass plants hold the soil?

3. Do you see signs of other living things besides grass?

4. **Think Critically** In your Science Journal, answer the above questions and describe any organisms that are present in your section of sod. Explain how these organisms might affect the growth of grass plants. Draw a picture of your section of sod.

Preview this chapter's content and activities at booke.msscience.com

Ecology Make the following Foldable to help organize information about one of your favorite wild animals and its role in an ecosystem.

STEP 1 **Fold** a vertical sheet of paper from side to side. Make the front edge 1.25 cm shorter than the back edge.

STEP 2 **Turn** lengthwise and **fold** into thirds.

STEP 3 **Unfold and cut** only the top layer along both folds to make three tabs. **Label** each tab.

Organism | Population | Community

Identify Questions Before you read the chapter, write what you already know about your favorite animal under the left tab of your Foldable. As you read the chapter, write how the animal is part of a population and a community under the appropriate tabs.

Compare and Contrast

1 Learn It! Good readers compare and contrast information as they read. This means they look for similarities and differences to help them to remember important ideas. Look for signal words in the text to let you know when the author is comparing or contrasting.

Compare and Contrast Signal Words	
Compare	**Contrast**
as	but
like	or
likewise	unlike
similarly	however
at the same time	although
in a similar way	on the other hand

2 Practice It! Read the excerpt below and notice how the author uses contrast signal words to describe the differences between the biotic potentials of species.

The highest rate of reproduction under ideal conditions is a population's biotic potential. The **larger** the number of offspring that are produced by parent organisms, the **higher** the biotic potential of the species will be. Compare an avocado tree to a tangerine tree.

—*from page 16*

3 Apply It! Compare and contrast the different types of symbiotic relationships on page 22.

Reading Tip

As you read, use other skills, such as summarizing and connecting, to help you understand comparisons and contrasts.

Target Your Reading

Use this to focus on the main ideas as you read the chapter.

1. **Before you read** the chapter, respond to the statements below on your worksheet or on a numbered sheet of paper.
   - Write an **A** if you **agree** with the statement.
   - Write a **D** if you **disagree** with the statement.

2. **After you read** the chapter, look back to this page to see if you've changed your mind about any of the statements.
   - If any of your answers changed, explain why.
   - Change any false statements into true statements.
   - Use your revised statements as a study guide.

Science Online
Print out a worksheet of this page at booke.msscience.com

Before You Read A or D		Statement	After You Read A or D
	1	A community is all the populations of species that live in an ecosystem.	
	2	All deserts are hot and dry environments.	
	3	An ecosystem is made up of only the living things in an area.	
	4	Organisms living in the wild always have enough food and living space.	
	5	The greatest competition in nature is among organisms of the same species.	
	6	Both nonliving and living parts of an ecosystem can limit the number of individuals in a population.	
	7	Living organisms do not need a constant supply of energy.	
	8	All consumers are predators.	
	9	Relationships between organisms of different species cannot benefit both organisms.	

Living Earth

as you read

What You'll Learn

- **Identify** places where life is found on Earth.
- **Define** ecology.
- **Observe** how the environment influences life.

Why It's Important

All living things on Earth depend on each other for survival.

🔍 Review Vocabulary

adaptation: any variation that makes an organism better suited to its environment

New Vocabulary

- ● biosphere
- ● ecosystem
- ● ecology
- ● population
- ● community
- ● habitat

The Biosphere

What makes Earth different from other planets in the solar system? One difference is Earth's abundance of living organisms. The part of Earth that supports life is the **biosphere** (BI uh sfihr). The biosphere includes the top portion of Earth's crust, all the waters that cover Earth's surface, and the atmosphere that surrounds Earth.

✔ **Reading Check** *What three things make up the biosphere?*

As **Figure 1** shows, the biosphere is made up of different environments that are home to different kinds of organisms. For example, desert environments receive little rain. Cactus plants, coyotes, and lizards are included in the life of the desert. Tropical rain forest environments receive plenty of rain and warm weather. Parrots, monkeys, and tens of thousands of other organisms live in the rain forest. Coral reefs form in warm, shallow ocean waters. Arctic regions near the north pole are covered with ice and snow. Polar bears, seals, and walruses live in the arctic.

Arctic

Desert

Coral reef

Figure 1 Earth's biosphere consists of many environments, including ocean waters, polar regions, and deserts.

Life on Earth In our solar system, Earth is the third planet from the Sun. The amount of energy that reaches Earth from the Sun helps make the temperature just right for life. Mercury, the planet closest to the Sun, is too hot during the day and too cold at night to make life possible there. Venus, the second planet from the Sun, has a thick, carbon dioxide atmosphere and high temperatures. It is unlikely that life could survive there. Mars, the fourth planet, is much colder than Earth because it is farther from the Sun and has a thinner atmosphere. It might support microscopic life, but none has been found. The planets beyond Mars probably do not receive enough heat and light from the Sun to have the right conditions for life.

Ecosystems

On a visit to Yellowstone National Park in Wyoming, you might see a prairie scene like the one shown in **Figure 2.** Bison graze on prairie grass. Cowbirds follow the bison, catching grasshoppers that jump away from the bisons' hooves. This scene is part of an ecosystem. An **ecosystem** consists of all the organisms living in an area, as well as the nonliving parts of that environment. Bison, grass, birds, and insects are living organisms of this prairie ecosystem. Water, temperature, sunlight, soil, and air are nonliving features of this prairie ecosystem. **Ecology** is the study of interactions that occur among organisms and their environments. Ecologists are scientists who study these interactions.

Reading Check *What is an ecosystem?*

Figure 2 Ecosystems are made up of living organisms and the nonliving factors of their environment. In this prairie ecosystem, cowbirds eat insects and bison graze on grass.
List *other kinds of organisms that might live in this ecosystem.*

Populations

Suppose you meet an ecologist who studies how a herd of bison moves from place to place and how the female bison in the herd care for their young. This ecologist is studying the members of a population. A **population** is made up of all organisms of the same species that live in an area at the same time. For example, all the bison in a prairie ecosystem are one population. All the cowbirds in this ecosystem make up a different population. The grasshoppers make up yet another population.

Ecologists often study how populations interact. For example, an ecologist might try to answer questions about several prairie species. How does grazing by bison affect the growth of prairie grass? How does grazing influence the insects that live in the grass and the birds that eat those insects? This ecologist is studying a community. A **community** is all the populations of all species living in an ecosystem. The prairie community is made of populations of bison, grasshoppers, cowbirds, and all other species in the prairie ecosystem. An arctic community might include populations of fish, seals that eat fish, and polar bears that hunt and eat seals. **Figure 3** shows how organisms, populations, communities, and ecosystems are related.

Figure 3 The living world is arranged in several levels of organization.

Figure 4 The trees of the forest provide a habitat for woodpeckers and other birds. This salamander's habitat is the moist forest floor.

Habitats

Each organism in an ecosystem needs a place to live. The place in which an organism lives is called its **habitat.** The animals shown in **Figure 4** live in a forest ecosystem. Trees are the woodpecker's habitat. These birds use their strong beaks to pry insects from tree bark or break open acorns and nuts. Woodpeckers usually nest in holes in dead trees. The salamander's habitat is the forest floor, beneath fallen leaves and twigs. Salamanders avoid sunlight and seek damp, dark places. This animal eats small worms, insects, and slugs. An organism's habitat provides the kinds of food and shelter, the temperature, and the amount of moisture the organism needs to survive.

section 1 review

Summary

The Biosphere

- The biosphere is the portion of Earth that supports life.

Ecosystems

- An ecosystem is made up of the living organisms and nonliving parts of an area.

Populations

- A population is made up of all members of a species that live in the same ecosystem.
- A community consists of all the populations in an ecosystem.

Habitats

- A habitat is where an organism lives.

Self Check

1. **List** three parts of the Earth included in the biosphere.
2. **Define** the term *ecology*.
3. **Compare and contrast** the terms *habitat* and *biosphere*.
4. **Identify** the major difference between a community and a population, and give one example of each.
5. **Think Critically** Does the amount of rain that falls in an area determine which kinds of organisms can live there? Why or why not?

Applying Skills

6. **Form a hypothesis** about how a population of dandelion plants might be affected by a population of rabbits.

Populations

What You'll Learn

- **Identify** methods for estimating population sizes.
- **Explain** how competition limits population growth.
- **List** factors that influence changes in population size.

Why It's Important

Competition caused by population growth reduces the amount of food, living space, and other resources available to organisms, including humans.

🔍 Review Vocabulary

natural selection: hypothesis that states organisms with traits best suited to their environment are more likely to survive and reproduce

New Vocabulary

- limiting factor
- carrying capacity

Competition

Wild crickets feed on plant material at night. They hide under leaves or in dark damp places during the day. In some pet shops, crickets are raised in cages and fed to pet reptiles. Crickets require plenty of food, water, and hiding places. As a population of caged crickets grows, extra food and more hiding places are needed. To avoid crowding, some crickets might have to be moved to other cages.

Food and Space Organisms living in the wild do not always have enough food or living space. The Gila woodpecker, shown in **Figure 5,** lives in the Sonoran Desert of Arizona and Mexico. This woodpecker makes its nest by drilling a hole in a saguaro (suh GWAR oh) cactus. Woodpeckers must compete with each other for nesting spots. Competition occurs when two or more organisms seek the same resource at the same time.

Growth Limits Competition limits population size. If available nesting spaces are limited, some woodpeckers will not be able to raise young. Gila woodpeckers eat cactus fruit, berries, and insects. If food becomes scarce, some woodpeckers might not survive to reproduce. Competition for food, living space, or other resources can limit population growth.

In nature, the most intense competition is usually among individuals of the same species, because they need the same kinds of food and shelter. Competition also takes place among different species. For example, after a Gila woodpecker has abandoned its nest, owls, flycatchers, snakes, and lizards might compete for the shelter of the empty hole.

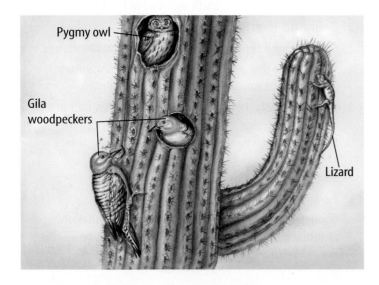

Pygmy owl

Gila woodpeckers

Lizard

Figure 5 Gila woodpeckers make nesting holes in the saguaro cactus. Many animals compete for the shelter these holes provide.

Population Size

Ecologists often need to measure the size of a population. This information can indicate whether or not a population is healthy and growing. Population counts can help identify populations that could be in danger of disappearing.

Some populations are easy to measure. If you were raising crickets, you could measure the size of your cricket population simply by counting all the crickets in the container. What if you wanted to compare the cricket populations in two different containers? You would calculate the number of crickets per square meter (m^2) of your container. The number of individuals of one species per a specific area is called population density. **Figure 6** shows Earth's human population density.

Reading Check *What is population density?*

Measuring Populations Counting crickets can be tricky. They look alike, move a lot, and hide. The same cricket could be counted more than once, and others could be completely missed. Ecologists have similar problems when measuring wildlife populations. One of the methods they use is called trap-mark-release. Suppose you want to count wild rabbits. Rabbits live underground and come out at dawn and dusk to eat. Ecologists set traps that capture rabbits without injuring them. Each captured rabbit is marked and released. Later, another sample of rabbits is captured. Some of these rabbits will have marks, but many will not. By comparing the number of marked and unmarked rabbits in the second sample, ecologists can estimate the population size.

Mini LAB

Observing Seedling Competition

Procedure

1. Fill **two plant pots** with **moist potting soil.**
2. Plant **radish seeds** in one pot, following the spacing instructions on the seed packet. Label this pot *Recommended Spacing*.
3. Plant **radish seeds** in the second pot, spaced half the recommended distance apart. Label this pot *Densely Populated*. Wash your hands.
4. Keep the soil moist. When the seeds sprout, move them to a well-lit area.
5. Measure and record in your **Science Journal** the height of the seedlings every two days for two weeks.

Analysis

1. Which plants grew faster?
2. Which plants looked healthiest after two weeks?
3. How did competition influence the plants?

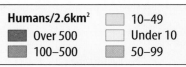

Try at Home

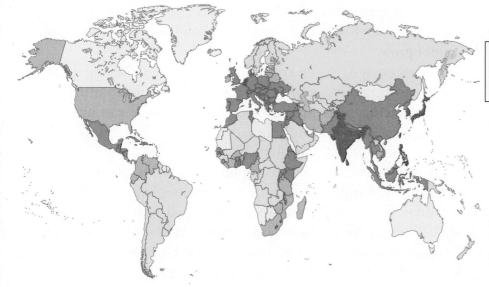

Humans/2.6km²	
Over 500	10–49
100–500	Under 10
	50–99

Figure 6 This map shows human population density. **Interpret Illustrations** *Which countries have the highest population density?*

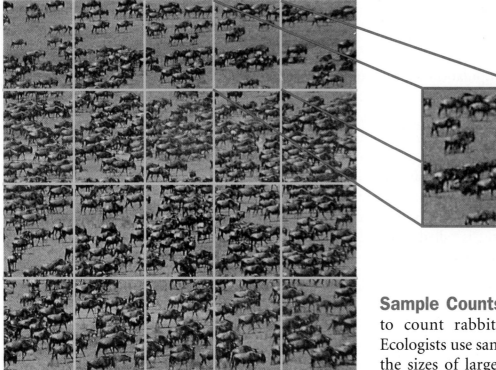

Figure 7 Ecologists can estimate population size by making a sample count. Wildebeests graze on the grassy plains of Africa.
Draw Conclusions *How could you use the enlarged square to estimate the number of wildebeests in the entire photograph?*

Sample Counts What if you wanted to count rabbits over a large area? Ecologists use sample counts to estimate the sizes of large populations. To estimate the number of rabbits in an area of 100 acres, for example, you could count the rabbits in one acre and multiply by 100 to estimate the population size. **Figure 7** shows another approach to sample counting.

Limiting Factors One grass plant can produce hundreds of seeds. Imagine those seeds drifting onto a vacant field. Many of the seeds sprout and grow into grass plants that produce hundreds more seeds. Soon the field is covered with grass. Can this grass population keep growing forever? Suppose the seeds of wildflowers or trees drift onto the field. If those seeds sprout, trees and flowers would compete with grasses for sunlight, soil, and water. Even if the grasses did not have to compete with other plants, they might eventually use up all the space in the field. When no more living space is available, the population cannot grow.

In any ecosystem, the availability of food, water, living space, mates, nesting sites, and other resources is often limited. A **limiting factor** is anything that restricts the number of individuals in a population. Limiting factors include living and nonliving features of the ecosystem.

A limiting factor can affect more than one population in a community. Suppose a lack of rain limits plant growth in a meadow. Fewer plants produce fewer seeds. For seed-eating mice, this reduction in the food supply could become a limiting factor. A smaller mouse population could, in turn, become a limiting factor for the hawks and owls that feed on mice.

Carrying Capacity A population of robins lives in a grove of trees in a park. Over several years, the number of robins increases and nesting space becomes scarce. Nesting space is a limiting factor that prevents the robin population from getting any larger. This ecosystem has reached its carrying capacity for robins. **Carrying capacity** is the largest number of individuals of one species that an ecosystem can support over time. If a population begins to exceed the environment's carrying capacity, some individuals will not have enough resources. They could die or be forced to move elsewhere, like the deer shown in **Figure 8.**

Figure 8 These deer might have moved into a residential area because a nearby forest's carrying capacity for deer has been reached.

✔ **Reading Check** *How are limiting factors related to carrying capacity?*

Applying Science

Do you have too many crickets?

You've decided to raise crickets to sell to pet stores. A friend says you should not allow the cricket population density to go over 210 crickets/m². Use what you've learned in this section to measure the population density in your cricket tanks.

Identifying the Problem

The table on the right lists the areas and populations of your three cricket tanks. How can you determine if too many crickets are in one tank? If a tank contains too many crickets, what could you do? Explain why too many crickets in a tank might be a problem.

Cricket Population

Tank	Area (m²)	Number of Crickets
1	0.80	200
2	0.80	150
3	1.5	315

Solving the Problem

1. Do any of the tanks contain too many crickets? Could you make the population density of the three tanks equal by moving crickets from one tank to another? If so, which tank would you move crickets into?

2. Wild crickets living in a field have a population density of 2.4 crickets/m². If the field's area is 250 m², what is the approximate size of the cricket population? Why would the population density of crickets in a field be lower than the population density of crickets in a tank?

Biotic Potential What would happen if no limiting factors restricted the growth of a population? Think about a population that has an unlimited supply of food, water, and living space. The climate is favorable. Population growth is not limited by diseases, predators, or competition with other species. Under ideal conditions like these, the population would continue to grow.

The highest rate of reproduction under ideal conditions is a population's biotic potential. The larger the number of offspring that are produced by parent organisms, the higher the biotic potential of the species will be. Compare an avocado tree to a tangerine tree. Assume that each tree produces the same number of fruits. Each avocado fruit contains one large seed. Each tangerine fruit contains a dozen seeds or more. Because the tangerine tree produces more seeds per fruit, it has a higher biotic potential than the avocado tree.

Changes in Populations

Birthrates and death rates also influence the size of a population and its rate of growth. A population gets larger when the number of individuals born is greater than the number of individuals that die. When the number of deaths is greater than the number of births, populations get smaller. Take the squirrels living in New York City's Central Park as an example. In one year, if 900 squirrels are born and 800 die, the population increases by 100. If 400 squirrels are born and 500 die, the population decreases by 100.

The same is true for human populations. **Table 1** shows birthrates, death rates, and population changes for several countries around the world. In countries with faster population growth, birthrates are much higher than death rates. In countries with slower population growth, birthrates are only slightly higher than death rates. In Germany, where the population is getting smaller, the birthrate is lower than the death rate.

Table 1 Population Growth			
	Birthrate*	Death Rate*	Population Increase (percent)
Rapid-Growth Countries			
Jordan	38.8	5.5	3.3
Uganda	50.8	21.8	2.9
Zimbabwe	34.3	9.4	5.2
Slow-Growth Countries			
Germany	9.4	10.8	−1.5
Sweden	10.8	10.6	0.1
United States	14.8	8.8	0.6

*Number per 1,000 people

Figure 9 Mangrove seeds sprout while they are still attached to the parent tree. Some sprouted seeds drop into the mud below the parent tree and continue to grow. Others drop into the water and can be carried away by tides and ocean currents. When they wash ashore, they might start a new population of mangroves or add to an existing mangrove population.

Moving Around Most animals can move easily from place to place, and these movements can affect population size. For example, a male mountain sheep might wander many miles in search of a mate. After he finds a mate, their offspring might establish a completely new population of mountain sheep far from the male's original population.

Many bird species move from one place to another during their annual migrations. During the summer, populations of Baltimore orioles are found throughout eastern North America. During the winter, these populations disappear because the birds migrate to Central America. They spend the winter there, where the climate is mild and food supplies are plentiful. When summer approaches, the orioles migrate back to North America.

Even plants and microscopic organisms can move from place to place, carried by wind, water, or animals. The tiny spores of mushrooms, mosses, and ferns float through the air. The seeds of dandelions, maple trees, and other plants have feathery or winglike growths that allow them to be carried by wind. Spine-covered seeds hitch rides by clinging to animal fur or people's clothing. Many kinds of seeds can be transported by river and ocean currents. Mangrove trees growing along Florida's Gulf Coast, shown in **Figure 9,** provide an example of how water moves seeds.

Mini LAB

Comparing Biotic Potential

Procedure

1. Remove all the seeds from a **whole fruit.** Do not put fruit or seeds in your mouth.
2. Count the total number of seeds in the fruit. Wash your hands, then record these data in your **Science Journal.**
3. Compare your seed totals with those of classmates who examined other types of fruit.

Analysis

1. Which type of fruit had the most seeds? Which had the fewest seeds?
2. What is an advantage of producing many seeds? Can you think of a possible disadvantage?
3. To estimate the total number of seeds produced by a tomato plant, what would you need to know?

Figure 10

When a species enters an ecosystem that has abundant food, water, and other resources, its population can flourish. Beginning with a few organisms, the population increases until the number of organisms and available resources are in balance. At that point, population growth slows or stops. A graph of these changes over time produces an S-curve, as shown here for coyotes.

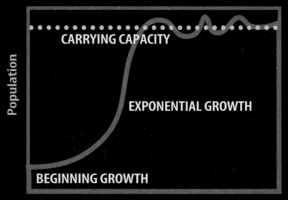

BEGINNING GROWTH During the first few years, population growth is slow, because there are few adults to produce young. As the population grows, so does the number of breeding adults.

EXPONENTIAL GROWTH As the number of adults in the population grows, so does the number of births. The coyote population undergoes exponential growth, quickly increasing in size.

CARRYING CAPACITY As resources become less plentiful, the birthrate declines and the death rate may rise. Population growth slows. The coyote population has reached the environmental carrying capacity—the maximum number of coyotes that the environment can sustain.

Exponential Growth When a species moves into a new area with plenty of food, living space, and other resources, the population grows quickly, in a pattern called exponential growth. Exponential growth means that the larger a population gets, the faster it grows. Over time, the population will reach the ecosystem's carrying capacity for that species. **Figure 10** shows each stage in this pattern of population growth.

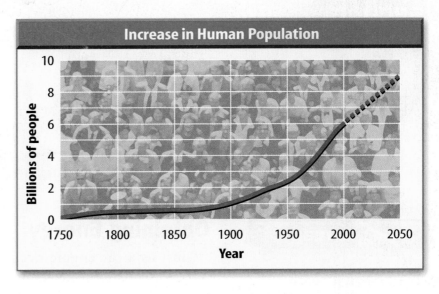

Increase in Human Population

As a population approaches its ecosystem's carrying capacity, competition for living space and other resources increases. As you can see in **Figure 11,** Earth's human population shows exponential growth. By the year 2050, the population could reach 9 billion. You probably have read about or experienced some of the competition associated with human population growth, such as freeway traffic jams, crowded subways and buses, or housing shortages. As population density increases, people are forced to live closer to one another. Infectious diseases can spread easily when people are crowded together.

Figure 11 The size of the human population is increasing by about 1.6 percent per year.
Identify *the factors that affect human population growth.*

section 2 review

Summary

Competition
- When more than one organism needs the same resource, competition occurs.
- Competition limits population size.

Population Size
- Population density is the number of individuals per unit area.
- Limiting factors are resources that restrict population size.
- An ecosystem's carrying capacity is the largest population it can support.
- Biotic potential is the highest possible rate of growth for a population.

Changes in Populations
- Birthrates, death rates, and movement from place to place affect population size.

Self Check

1. **Describe** three ways in which ecologists can estimate the size of a population.
2. **Explain** how birthrates and death rates influence the size of a population.
3. **Explain** how carrying capacity influences the number of organisms in an ecosystem.
4. **Think Critically** Why are food and water the limiting factors that usually have the greatest effect on population size?

Applying Skills

5. **Make and use a table** on changes in the size of a deer population in Arizona. Use the following data. In 1910 there were 6 deer; in 1915, 36 deer; in 1920, 143 deer; in 1925, 86 deer; and in 1935, 26 deer. Explain what might have caused these changes.

Interactions Within Communities

What **You'll Learn**

- **Describe** how organisms obtain energy for life.
- **Explain** how organisms interact.
- **Recognize** that every organism occupies a niche.

Why **It's Important**

Obtaining food, shelter, and other needs is crucial to the survival of all living organisms, including you.

🔍 **Review Vocabulary**

social behavior: interactions among members of the same species

New Vocabulary

- producer
- consumer
- symbiosis
- mutualism
- commensalism
- parasitism
- niche

Obtaining Energy

Just as a car engine needs a constant supply of gasoline, living organisms need a constant supply of energy. The energy that fuels most life on Earth comes from the Sun. Some organisms use the Sun's energy to create energy-rich molecules through the process of photosynthesis. The energy-rich molecules, usually sugars, serve as food. They are made up of different combinations of carbon, hydrogen, and oxygen atoms. Energy is stored in the chemical bonds that hold the atoms of these molecules together. When the molecules break apart—for example, during digestion—the energy in the chemical bonds is released to fuel life processes.

Producers Organisms that use an outside energy source like the Sun to make energy-rich molecules are called **producers.** Most producers contain chlorophyll (KLOR uh fihl), a chemical that is required for photosynthesis. As shown in **Figure 12,** green plants are producers. Some producers do not contain chlorophyll and do not use energy from the Sun. Instead, they make energy-rich molecules through a process called chemosynthesis (kee moh SIHN thuh sus). These organisms can be found near volcanic vents on the ocean floor. Inorganic molecules in the water provide the energy source for chemosynthesis.

Euglena
LM Magnification: 125×

Algae
LM Magnification: 25×

Figure 12 Green plants, including the grasses that surround this pond, are producers. The pond water also contains producers, including microscopic organisms like *Euglena* and algae.

Figure 13 Four categories of consumers are shown.
Identify *the consumer category that would apply to a bear. What about a mushroom?*

Herbivores	Carnivores	Omnivores	Decomposers

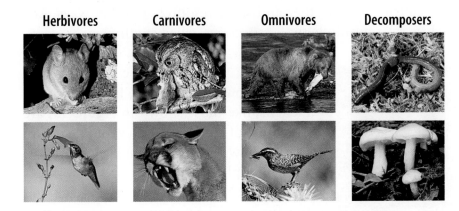

Consumers

Organisms that cannot make their own energy-rich molecules are called **consumers.** Consumers obtain energy by eating other organisms. **Figure 13** shows the four general categories of consumers. Herbivores are the vegetarians of the world. They include rabbits, deer, and other plant eaters. Carnivores are animals that eat other animals. Frogs and spiders are carnivores that eat insects. Omnivores, including pigs and humans, eat mostly plants and animals. Decomposers, including fungi, bacteria, and earthworms, consume wastes and dead organisms. Decomposers help recycle once-living matter by breaking it down into simple, energy-rich substances. These substances might serve as food for decomposers, be absorbed by plant roots, or be consumed by other organisms.

Reading Check *How are producers different from consumers?*

Food Chains

Ecology includes the study of how organisms depend on each other for food. A food chain is a simple model of the feeding relationships in an ecosystem. For example, shrubs are food for deer, and deer are food for mountain lions, as illustrated in **Figure 14.** What food chain would include you?

INTEGRATE Chemistry

Glucose The nutrient molecule produced during photosynthesis is glucose. Look up the chemical structure of glucose and draw it in your Science Journal.

Figure 14 Food chains illustrate how consumers obtain energy from other organisms in an ecosystem.

Symbiotic Relationships

Figure 15 Many examples of symbiotic relationships exist in nature.

Not all relationships among organisms involve food. Many organisms live together and share resources in other ways. Any close relationship between species is called **symbiosis.**

Lichens are a result of mutualism.

Mutualism You may have noticed crusty lichens growing on fences, trees, or rocks. Lichens, like those shown in **Figure 15,** are made up of an alga or a cyanobacterium that lives within the tissues of a fungus. Through photosynthesis, the cyanobacterium or alga supplies energy to itself and the fungus. The fungus provides a protected space in which the cyanobacterium or alga can live. Both organisms benefit from this association. A symbiotic relationship in which both species benefit is called **mutualism** (MYEW chuh wuh lih zum).

Clown fish and sea anemones have a commensal relationship.

LM Magnification: 128×

Commensalism If you've ever visited a marine aquarium, you might have seen the ocean organisms shown in **Figure 15.** The creature with gently waving, tubelike tentacles is a sea anemone. The tentacles contain a mild poison. Anemones use their tentacles to capture shrimp, fish, and other small animals to eat. The striped clown fish can swim among the tentacles without being harmed. The anemone's tentacles protect the clown fish from predators. In this relationship, the clown fish benefits but the sea anemone is not helped or hurt. A symbiotic relationship in which one organism benefits and the other is not affected is called **commensalism** (kuh MEN suh lih zum).

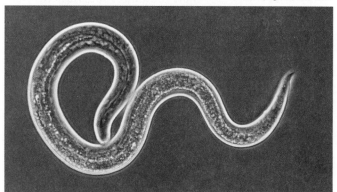

Some roundworms are parasites that rob nutrients from their hosts.

Parasitism Pet cats or dogs sometimes have to be treated for worms. Roundworms, like the one shown in **Figure 15,** are common in puppies. This roundworm attaches itself to the inside of the puppy's intestine and feeds on nutrients in the puppy's blood. The puppy may have abdominal pain, bloating, and diarrhea. If the infection is severe, the puppy might die. A symbiotic relationship in which one organism benefits but the other is harmed is called **parasitism** (PER uh suh tih zum).

Niches

One habitat might contain hundreds or even thousands of species. Look at the rotting log habitat shown in **Figure 16.** A rotting log in a forest can be home to many species of insects, including termites that eat decaying wood and ants that feed on the termites. Other species that live on or under the rotting log include millipedes, centipedes, spiders, and worms. You might think that competition for resources would make it impossible for so many species to live in the same habitat. However, each species has different requirements for its survival. As a result, each species has its own niche (NICH). An organism's **niche** is its role in its environment—how it obtains food and shelter, finds a mate, cares for its young, and avoids danger.

✓ Reading Check *Why does each species have its own niche?*

Special adaptations that improve survival are often part of an organism's niche. Milkweed plants contain a poison that prevents many insects from feeding on them. Monarch butterfly caterpillars have an adaptation that allows them to eat milkweed. Monarchs can take advantage of a food resource that other species cannot use. Milkweed poison also helps protect monarchs from predators. When the caterpillars eat milkweed, they become slightly poisonous. Birds avoid eating monarchs because they learn that the caterpillars and adult butterflies have an awful taste and can make them sick.

INTEGRATE History

Plant Poisons The poison in milkweed is similar to the drug digitalis. Small amounts of digitalis are used to treat heart ailments in humans, but it is poisonous in large doses. Research the history of digitalis as a medicine. In your Science Journal, list diseases for which it was used but is no longer used.

Figure 16 Different adaptations enable each species living in this rotting log to have its own niche. Termites eat wood. They make tunnels inside the log. Millipedes feed on plant matter and find shelter beneath the log. Wolf spiders capture insects living in and around the log.

Termites

Millipede

Wolf spider

Predator and Prey When you think of survival in the wild, you might imagine an antelope running away from a lion. An organism's niche includes how it avoids being eaten and how it finds or captures its food. Predators, like the one shown in **Figure 17,** are consumers that capture and eat other consumers. The prey is the organism that is captured by the predator. The presence of predators usually increases the number of different species that can live in an ecosystem. Predators limit the size of prey populations. As a result, food and other resources are less likely to become scarce, and competition between species is reduced.

Figure 17 The alligator is a predator. The turtle is its prey.

Cooperation Individual organisms often cooperate in ways that improve survival. For example, a white-tailed deer that detects the presence of wolves or coyotes will alert the other deer in the herd. Many insects, such as ants and honeybees, live in social groups. Different individuals perform different tasks required for the survival of the entire nest. Soldier ants protect workers that go out of the nest to gather food. Worker ants feed and care for ant larvae that hatch from eggs laid by the queen. These cooperative actions improve survival and are a part of the specie's niche.

section 3 review

Summary

Obtaining Energy
- All life requires a constant supply of energy.
- Most producers make food by photosynthesis using light energy.
- Consumers cannot make food. They obtain energy by eating producers or other consumers.
- A food chain models the feeding relationships between species.

Symbiotic Relationships
- Symbiosis is any close relationship between species.
- Mutualism, commensalism, and parasitism are types of symbiosis.
- An organism's niche describes the ways in which the organism obtains food, avoids danger, and finds shelter.

Self Check

1. **Explain** why all consumers depend on producers for food.
2. **Describe** a mutualistic relationship between two imaginary organisms. Name the organisms and explain how each benefits.
3. **Compare and contrast** the terms *habitat* and *niche.*
4. **Think Critically** A parasite can obtain food only from a host organism. Explain why most parasites weaken, but do not kill, their hosts.

Applying Skills

5. **Design an experiment** to classify the symbiotic relationship that exists between two hypothetical organisms. Animal A definitely benefits from its relationship with Plant B, but it is not clear whether Plant B benefits, is harmed, or is unaffected.

Feeding Habits of Planaria

You probably have watched minnows darting about in a stream. It is not as easy to observe organisms that live at the bottom of a stream, beneath rocks, logs, and dead leaves. Countless stream organisms, including insect larvae, worms, and microscopic organisms, live out of your view. One such organism is a type of flatworm called a planarian. In this lab, you will find out about the eating habits of planarians.

◉ Real-World Question

What food items do planarians prefer to eat?

Goals
- **Observe** the food preference of planarians.
- **Infer** what planarians eat in the wild.

Materials
small bowl   guppies (several)
planarians (several)   pond or stream water
lettuce leaf   magnifying lens
raw liver or meat

Safety Precautions

◉ Procedure

1. Fill the bowl with stream water.
2. Place a lettuce leaf, piece of raw liver, and several guppies in the bowl. Add the planarians. Wash your hands.
3. **Observe** what happens inside the bowl for at least 20 minutes. Do not disturb the bowl or its contents. Use a magnifying lens to look at the planarians.
4. **Record** all of your observations in your Science Journal.

◉ Conclude and Apply

1. **Name** the food the planarians preferred.
2. **Infer** what planarians might eat when in their natural environment.
3. **Describe,** based on your observations during this lab, a planarian's niche in a stream ecosystem.
4. **Predict** where in a stream you might find planarians. Use references to find out whether your prediction is correct.

Magnification: Unknown

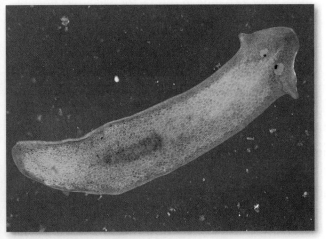

𝒞 ommunicating
Your Data

Share your results with other students in your class. Plan an adult-supervised trip with several classmates to a local stream to search for planarians in their native habitat. **For more help, refer to the** Science Skill Handbook.

LAB Design Your Own

POPULATION GROWTH IN FRUIT FLIES

Goals

- **Identify** the environmental factors needed by a population of fruit flies.
- **Design** an experiment to investigate how a change in one environmental factor affects in any way the size of a fruit fly population.
- **Observe** and **measure** changes in population size.

Possible Materials

fruit flies
standard fruit fly culture kit
food items (banana, orange peel, or other fruit)
water
heating or cooling source
culture containers
cloth, plastic, or other tops for culture containers
magnifying lens

Safety Precautions

⊙ Real-World Question

Populations can grow at an exponential rate only if the environment provides the right amount of food, shelter, air, moisture, heat, living space, and other factors. You probably have seen fruit flies hovering near ripe bananas or other fruit. Fruit flies are fast-growing organisms often raised in science laboratories. The flies are kept in culture tubes and fed a diet of specially prepared food flakes. Can you improve on this standard growing method to achieve faster population growth? Will a change in one environmental factor affect the growth of a fruit fly population?

⊙ Form a Hypothesis

Based on your reading about fruit flies, state a hypothesis about how changing one environmental factor will affect the rate of growth of a fruit fly population.

⊙ Test Your Hypothesis

Make a Plan

1. As a group, decide on one environmental factor to investigate. Agree on a hypothesis about how a change in this factor will affect population growth. Decide how you will test your hypothesis, and identify the experimental results that would support your hypothesis.

2. **List** the steps you will need to take to test your hypothesis. Describe exactly what you will do. List your materials.

3. **Determine** the method you will use to measure changes in the size of your fruit fly populations.

4. Prepare a data table in your Science Journal to record weekly measurements of your fruit fly populations.

5. Read the entire experiment and make sure all of the steps are in a logical order.

6. **Research** the standard method used to raise fruit flies in the laboratory. Use this method as the control in your experiment.

7. **Identify** all constants, variables, and controls in your experiment.

Follow Your Plan

1. Make sure your teacher approves your plan before you start.

2. Carry out your experiment.

3. **Measure** the growth of your fruit fly populations weekly and record the data in your data table.

◉ *Analyze Your Data*

1. **Identify** the constants and the variables in your experiment.

2. **Compare** changes in the size of your control population with changes in your experimental population. Which population grew faster?

3. **Make and Use Graphs** Using the information in your data table, make a line graph that shows how the sizes of your two fruit fly populations changed over time. Use a different colored pencil for each population's line on the graph.

◉ *Conclude and Apply*

1. **Explain** whether or not the results support your hypothesis.

2. **Compare** the growth of your control and experimental populations. Did either population reach exponential growth?
How do you know?

*C*ommunicating
Your Data

Compare the results of your experiment with those of other students in your class.
For more help, refer to the Science Skill Handbook.

The Census measures a human population

Counting people is important to the United States and to many other countries around the world. It helps governments determine the distribution of people in the various regions of a nation. To obtain this information, the government takes a census— a count of how many people are living in their country on a particular day at a particular time, and in a particular place. A census is a snapshot of a country's population.

Counting on the Count

When the United States government was formed, its founders set up the House of Representatives based on population. Areas with more people had more government representatives, and areas with fewer people had fewer representatives. In 1787, the requirement for a census became part of the U.S. Constitution. A census must be taken every ten years so the proper number of representatives for each state can be calculated.

The Short Form

Before 1970, United States census data was collected by field workers. They went door to door to count the number of people living in each household. Since then, the census has been done mostly by mail. Census data are important in deciding how to distribute government services and funding.

The 2000 Snapshot

One of the findings of the 2000 Census is that the U.S. population is becoming more equally spread out across age groups. Census officials estimate that by 2020 the population of children, middle-aged people, and senior citizens will be about equal. It's predicted also that there will be more people who are over 100 years old than ever before. Federal, state, and local governments will be using the results of the 2000 Census for years to come as they plan our future.

Census Develop a school census. What questions will you ask? (Don't ask questions that are too personal.) Who will ask them? How will you make sure you counted everyone? Using the results, can you make any predictions about your school's future or its current students?

Science Online

For more information, visit
booke.msscience.com/time

Reviewing Main Ideas

Section 1 Living Earth

1. Ecology is the study of interactions that take place in the biosphere.

2. A population is made up of all organisms of one species living in an area at the same time.

3. A community is made up of all the populations living in one ecosystem.

4. Living and nonliving factors affect an organism's ability to survive in its habitat.

Section 2 Populations

1. Population size can be estimated by counting a sample of a total population.

2. Competition for limiting factors can restrict the size of a population.

3. Population growth is affected by birthrate, death rate, and the movement of individuals into or out of a community.

4. Exponential population growth can occur in environments that provide a species with plenty of food, shelter, and other resources.

Section 3 Interactions Within Communities

1. All life requires energy.

2. Most producers use light to make food in the form of energy-rich molecules. Consumers obtain energy by eating other organisms.

3. Mutualism, commensalism, and parasitism are the three kinds of symbiosis.

4. Every species has its own niche, which includes adaptations for survival.

Visualizing Main Ideas

Copy and complete the following concept map on communities.

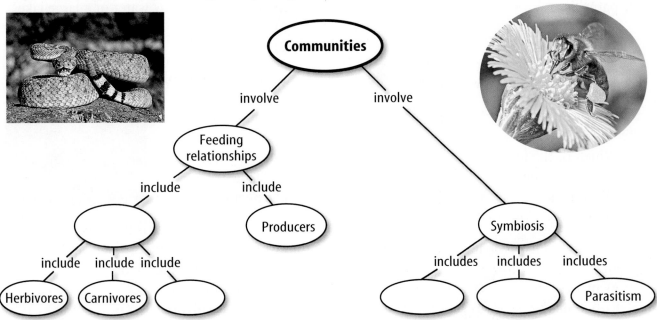

Using Vocabulary

biosphere p. 8	limiting factor p. 14
carrying capacity p. 15	mutualism p. 22
commensalism p. 22	niche p. 23
community p. 10	parasitism p. 22
consumer p. 21	population p. 20
ecology p. 19	producer p. 20
ecosystem p. 19	symbiosis p. 22
habitat p. 21	

Explain the difference between the vocabulary words in each of the following sets.

1. niche—habitat

2. mutualism—commensalism

3. limiting factor—carrying capacity

4. biosphere—ecosystem

5. producer—consumer

6. population—ecosystem

7. community—population

8. parasitism—symbiosis

9. ecosystem—ecology

10. parasitism—commensalism

Checking Concepts

Choose the word or phrase that best answers the question.

11. Which of the following is a living factor in the environment?
    A) animals   C) sunlight
    B) air   D) soil

12. What is made up of all the populations in an area?
    A) niches   C) community
    B) habitats   D) ecosystem

13. What does the number of individuals in a population that occupies an area of a specific size describe?
    A) clumping   C) spacing
    B) size   D) density

14. Which of the following animals is an example of an herbivore?
    A) wolf   C) tree
    B) moss   D) rabbit

15. What term best describes a symbiotic relationship in which one species is helped and the other is harmed?
    A) mutualism   C) commensalism
    B) parasitism   D) consumerism

16. Which of the following conditions tends to increase the size of a population?
    A) births exceed deaths
    B) population size exceeds the carrying capacity
    C) movements out of an area exceed movements into the area
    D) severe drought

17. Which of the following is most likely to be a limiting factor in a population of fish living in the shallow water of a large lake?
    A) sunlight   C) food
    B) water   D) soil

18. In which of the following categories does the pictured organism belong?
    A) herbivore
    B) carnivore
    C) producer
    D) consumer

19. Which pair of words is incorrect?
    A) black bear—carnivore
    B) grasshopper—herbivore
    C) pig—omnivore
    D) lion—carnivore

Science Online booke.msscience.com/vocabulary_puzzlemaker

Thinking Critically

20. Infer why a parasite has a harmful effect on the organism it infects.

21. Explain what factors affect carrying capacity.

22. Describe your own habitat and niche.

23. Make and Use Tables Copy and complete the following table.

Types of Symbiosis		
Organism A	**Organism B**	**Relationship**
Gains	Doesn't gain or lose	
Gains		Mutualism
Gains	Loses	

24. Explain how several different niches can exist in the same habitat.

25. Make a model of a food chain using the following organisms: grass, snake, mouse, and hawk.

26. Predict Dandelion seeds can float great distances on the wind with the help of white, featherlike attachments. Predict how a dandelion seed's ability to be carried on the wind helps reduce competition among dandelion plants.

27. Classify the following relationships as parasitism, commensalism, or mutualism: a shark and a remora fish that cleans and eats parasites from the shark's gills; head lice and a human; a spiny sea urchin and a tiny fish that hides from predators by floating among the sea urchin's spines.

28. Compare and contrast the diets of omnivores and herbivores. Give examples of each.

29. List three ways exponential growth in the human population affects people's lives.

Performance Activities

30. Poster Use photographs from old magazines to create a poster that shows at least three different food chains. Illustrate energy pathways from organism to organism and from organisms to the environment. Display your poster for your classmates.

Applying Math

31. Measuring Populations An ecologist wants to know the size of a population of wild daisy plants growing in a meadow that measures 1,000 m². The ecologist counts 30 daisy plants in a sample area of 100 m². What is the estimated population of daisies in the entire meadow?

Use the table below to answer question 32.

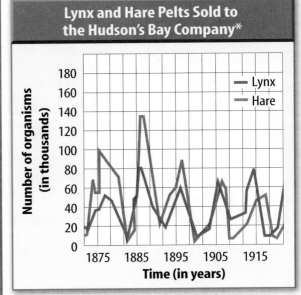

Lynx and Hare Pelts Sold to the Hudson's Bay Company*

* Data from 1875 through 1904 reflects actual pelts counted. Data from 1905 through 1915 is based on answers to questionnaire.

32. Changes in Populations The graph above shows changes over time in the sizes of lynx and rabbit populations in an ecosystem. What does the graph tell you about the relationship between these two species? Explain how they influence each other's population size.

Part 1 Multiple Choice

Record your answers on the answer sheet provided by your teacher or on a sheet of paper.

1. Which of the following terms is defined in part by nonliving factors?
   A. population
   B. community
   C. ecosystem
   D. niche

2. Which of the follow terms would include all places where organisms live on Earth?
   A. ecosystem
   B. habitat
   C. biosphere
   D. community

3. Which of the following is not a method of measuring populations?
   A. total count
   B. trap-release
   C. sample count
   D. trap-mark-release

Use the photo below to answer questions 4 and 5.

4. Dead plants at the bottom of this pond are consumed by
   A. omnivores.
   B. herbivores.
   C. carnivores.
   D. decomposers.

5. If the pond shrinks in size, what effect will this have on the population density of the pond's minnow species?
   A. It will increase.
   B. It will decrease.
   C. It will stay the same.
   D. No effect; it is not a limiting factor.

6. Which of the following includes organisms that can directly convert energy from the Sun into food?
   A. producers
   B. decomposers
   C. omnivores
   D. consumers

7. You have a symbiotic relationship with bacteria in your digestive system. These bacteria break down food you ingest, and you get vital nutrients from them. Which type of symbiosis is this?
   A. mutualism
   B. barbarism
   C. commensalism
   D. parasitism

Use the photo below to answer questions 8 and 9.

8. An eastern screech owl might compete with which organism most intensely for resources?
   A. mouse
   B. hawk
   C. mountain lion
   D. wren

9. Which of the following organisms might compete with the mouse for seeds?
   A. hawk
   B. lion
   C. fox
   D. sparrow

10. Which of the following is an example of a community?
    A. all the white-tailed deer in a forest
    B. all the trees, soil, and water in a forest
    C. all the plants and animals in a wetland
    D. all the cattails in a wetland

Part 2 | Short Response/Grid In

Record your answers on the answer sheet provided by your teacher or on a sheet of paper.

Use the graph below to answer question 11.

Mouse Population Exposed to Predators

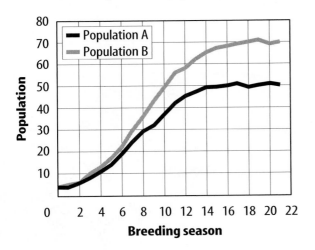

- Population A
- Population B

Population (y-axis): 10, 20, 30, 40, 50, 60, 70, 80

Breeding season (x-axis): 0, 2, 4, 6, 8, 10, 12, 14, 16, 18, 20, 22

11. The graph depicts the growth of two white-footed mice populations, one exposed to hawks (population A) and one without hawks (population B). Are hawks a limiting factor for either mouse population? If not, then what other factor could be a limiting factor for that population?

12. Diagram the flow of energy through an ecosystem. Include the sources of energy, producers, consumers, and decomposers in the ecosystem.

Test-Taking Tip

Understand the Question Be sure you understand the question before you read the answer choices. Make special note of words like NOT or EXCEPT. Read and consider choices before you mark your answer sheet.

Question 11 Make sure you understand which mouse population is subject to predation by hawks and which mouse population do hawks not affect.

Part 3 | Open Ended

Record your answers on a sheet of paper.

13. The colors and patterns of the viceroy butterfly are similar to the monarch butterfly, however, the viceroy caterpillars don't feed on milkweed. How does the viceroy butterfly benefit from this adaptation of its appearance? Under what circumstance would this adaptation not benefit the viceroy? Why?

Use the illustration below to answer question 14.

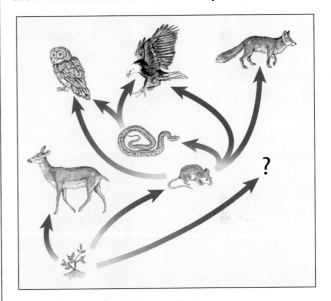

14. The illustration depicts a food web for a particular ecosystem. If the "?" is another mouse species population that is introduced into the ecosystem, explain what impact this would have on the species populations in the ecosystem.

15. Identify and explain possible limiting factors that would control the size of an ant colony.

16. How would you measure the size of a population of gray squirrels in a woodland? Explain which method you would choose and why.

The Nonliving Environment

The BIG Idea

Environments have both living and nonliving elements.

SECTION 1
Abiotic Factors
Main Idea Both living and nonliving parts of an environment are needed for organisms to survive.

SECTION 2
Cycles in Nature
Main Idea Many nonliving elements on Earth, such as water and oxygen, are recycled over and over.

SECTION 3
Energy Flow
Main Idea All living things use energy.

Sun, Surf, and Sand

Living things on this coast directly or indirectly depend on nonliving things, such as sunlight, water, and rocks, for energy and raw materials needed for their life processes. In this chapter, you will read how these and other nonliving things affect life on Earth.

Science Journal List all the nonliving things that you can see in this picture in order of importance. Explain your reasoning for the order you chose.

Start-Up Activities

Earth Has Many Ecosystems

Do you live in a dry, sandy region covered with cactus plants or desert scrub? Is your home in the mountains? Does snow fall during the winter? In this chapter, you'll learn why the nonliving factors in each ecosystem are different. The following lab will get you started.

1. Locate your city or town on a globe or world map. Find your latitude. Latitude shows your distance from the equator and is expressed in degrees, minutes, and seconds.

2. Locate another city with the same latitude as your city but on a different continent.

3. Locate a third city with latitude close to the equator.

4. Using references, compare average annual precipitation and average high and low temperatures for all three cities.

5. **Think Critically** Hypothesize how latitude affects average temperatures and rainfall.

Preview this chapter's content and activities at
booke.msscience.com

Nonliving Factors Make the following Foldable to help you understand the cause and effect relationships within the nonliving environment.

STEP 1 Fold two vertical sheets of paper in half from top to bottom. Cut the papers in half along the folds.

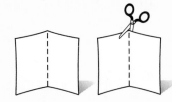

STEP 2 Discard one piece and fold the three vertical pieces in half from top to bottom.

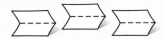

STEP 3 Turn the papers horizontally. Tape the short ends of the pieces together (overlapping the edges slightly).

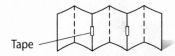

Tape

STEP 4 On one side, label the folds: *Nonliving, Water, Soil, Wind, Temperature,* and *Elevation.* Draw a picture of a familiar ecosystem on the other side.

Sequence As you read the chapter, write on the folds how each nonliving factor affects the environment that you draw.

Get Ready to Read

Make Inferences

1 Learn It! When you make inferences, you draw conclusions that are not directly stated in the text. This means you "read between the lines." You interpret clues and draw upon prior knowledge. Authors rely on a reader's ability to infer because all the details are not always given.

2 Practice It! Read the excerpt below and pay attention to highlighted words as you make inferences. Use this Think-Through chart to help you make inferences.

Water vapor that has been released into the atmosphere eventually comes into contact with colder air. The temperature of the water vapor drops. Over time, the water vapor cools enough to change back into liquid water. The process of changing from a gas to a liquid is called condensation. Water vapor condenses on particles of dust in the air, forming tiny droplets. At first, the droplets clump together to form clouds.

—*from page 45*

Text	Question	Inferences
Released into the atmosphere	How is the water vapor released back into the atmosphere?	Evaporation? Animals?
Comes into contact with colder air	Where does it come into contact with cold air?	High in the atmosphere? Cool air from cool weather?
Condenses on particles of dust	Where have the dust particles come from?	Pollution? Wind?

3 Apply It! As you read this chapter, practice your skill at making inferences by making connections and asking questions.

Reading Tip

Sometimes you make inferences by using other reading skills, such as questioning and predicting.

Target Your Reading

Use this to focus on the main ideas as you read the chapter.

① **Before you read** the chapter, respond to the statements below on your worksheet or on a numbered sheet of paper.
  - Write an **A** if you **agree** with the statement.
  - Write a **D** if you **disagree** with the statement.

② **After you read** the chapter, look back to this page to see if you've changed your mind about any of the statements.
  - If any of your answers changed, explain why.
  - Change any false statements into true statements.
  - Use your revised statements as a study guide.

Science Online
Print out a worksheet of this page at booke.msscience.com

Before You Read A or D	Statement		After You Read A or D
	1	The nonliving part on an environment often determines what living organisms are found there.	
	2	Most living organisms are mainly made up of water.	
	3	Heat from the Sun is responsible for wind.	
	4	Animals do not release water vapor.	
	5	The air we breathe mostly contains nitrogen.	
	6	Photosynthesis uses oxygen to produce energy.	
	7	Energy can be both converted to other forms and recycled.	
	8	Matter can be converted to other forms, but cannot be recycled.	
	9	The majority of energy is found at the bottom on an energy pyramid.	
	10	Water is a living part of the environment.	

Abiotic Factors

as you read

What You'll Learn

- **Identify** common abiotic factors in most ecosystems.
- **List** the components of air that are needed for life.
- **Explain** how climate influences life in an ecosystem.

Why It's Important

Knowing how organisms depend on the nonliving world can help humans maintain a healthy environment.

🔎 Review Vocabulary

environment: everything, such as climate, soil, and living things, that surrounds and affects an organism

New Vocabulary

- biotic
- abiotic
- atmosphere
- soil
- climate

Environmental Factors

Living organisms depend on one another for food and shelter. The leaves of plants provide food and a home for grasshoppers, caterpillars, and other insects. Many birds depend on insects for food. Dead plants and animals decay and become part of the soil. The features of the environment that are alive, or were once alive, are called **biotic** (bi AH tihk) factors. The term *biotic* means "living."

Biotic factors are not the only things in an environment that are important to life. Most plants cannot grow without sunlight, air, water, and soil. Animals cannot survive without air, water, or the warmth that sunlight provides. The nonliving, physical features of the environment are called **abiotic** (ay bi AH tihk) factors. The prefix *a* means "not." The term *abiotic* means "not living." Abiotic factors include air, water, soil, sunlight, temperature, and climate. The abiotic factors in an environment often determine which kinds of organisms can live there. For example, water is an important abiotic factor in the environment, as shown in **Figure 1.**

Figure 1 Abiotic factors—air, water, soil, sunlight, temperature, and climate—influence all life on Earth.

Air

Air is invisible and plentiful, so it is easily overlooked as an abiotic factor of the environment. The air that surrounds Earth is called the **atmosphere.** Air contains 78 percent nitrogen, 21 percent oxygen, 0.94 percent argon, 0.03 percent carbon dioxide, and trace amounts of other gases. Some of these gases provide substances that support life.

Carbon dioxide (CO_2) is required for photosynthesis. Photosynthesis—a series of chemical reactions—uses CO_2, water, and energy from sunlight to produce sugar molecules. Organisms, like plants, that can use photosynthesis are called producers because they produce their own food. During photosynthesis, oxygen is released into the atmosphere.

When a candle burns, oxygen from the air chemically combines with the molecules of candle wax. Chemical energy stored in the wax is converted and released as heat and light energy. In a similar way, cells use oxygen to release the chemical energy stored in sugar molecules. This process is called respiration. Through respiration, cells obtain the energy needed for all life processes. Air-breathing animals aren't the only organisms that need oxygen. Plants, some bacteria, algae, fish, and other organisms need oxygen for respiration.

Water

Water is essential to life on Earth. It is a major ingredient of the fluid inside the cells of all organisms. In fact, most organisms are 50 percent to 95 percent water. Respiration, digestion, photosynthesis, and many other important life processes can take place only in the presence of water. As **Figure 2** shows, environments that have plenty of water usually support a greater diversity of and a larger number of organisms than environments that have little water.

Figure 2 Water is an important abiotic factor in deserts and rain forests.

Life in deserts is limited to species that can survive for long periods without water.

Thousands of species can live in lush rain forests where rain falls almost every day.

Soil

Soil is a mixture of mineral and rock particles, the remains of dead organisms, water, and air. It is the topmost layer of Earth's crust, and it supports plant growth. Soil is formed, in part, of rock that has been broken down into tiny particles.

Soil is considered an abiotic factor because most of it is made up of nonliving rock and mineral particles. However, soil also contains living organisms and the decaying remains of dead organisms. Soil life includes bacteria, fungi, insects, and worms. The decaying matter found in soil is called humus. Soils contain different combinations of sand, clay, and humus. The type of soil present in a region has an important influence on the kinds of plant life that grow there.

Sunlight

All life requires energy, and sunlight is the energy source for almost all life on Earth. During photosynthesis, producers convert light energy into chemical energy that is stored in sugar molecules. Consumers are organisms that cannot make their own food. Energy is passed to consumers when they eat producers or other consumers. As shown in **Figure 3,** photosynthesis cannot take place if light is never available.

Shady forest

Bottom of deep ocean

Figure 3 Photosynthesis requires light. Little sunlight reaches the shady forest floor, so plant growth beneath trees is limited. Sunlight does not reach into deep lake or ocean waters. Photosynthesis can take place only in shallow water or near the water's surface.

Infer *how fish that live at the bottom of the deep ocean obtain energy.*

Figure 4 Temperature is an abiotic factor that can affect an organism's survival.

The penguin has a thick layer of fat to hold in heat and keep the bird from freezing. These emperor penguins huddle together for added warmth.

The Arabian camel stores fat only in its hump. This way, the camel loses heat from other parts of its body, which helps it stay cool in the hot desert.

Temperature

Sunlight supplies life on Earth with light energy for photosynthesis and heat energy for warmth. Most organisms can survive only if their body temperatures stay within the range of 0°C to 50°C. Water freezes at 0°C. The penguins in **Figure 4** are adapted for survival in the freezing Antarctic. Camels can survive the hot temperatures of the Arabian Desert because their bodies are adapted for staying cool. The temperature of a region depends in part on the amount of sunlight it receives. The amount of sunlight depends on the land's latitude and elevation.

Figure 5 Because Earth is curved, latitudes farther from the equator are colder than latitudes near the equator.

✔ **Reading Check**   *What does sunlight provide for life on Earth?*

Latitude In this chapter's Launch Lab, you discovered that temperature is affected by latitude. You found that cities located at latitudes farther from the equator tend to have colder temperatures than cities at latitudes nearer to the equator. As **Figure 5** shows, polar regions receive less of the Sun's energy than equatorial regions. Near the equator, sunlight strikes Earth directly. Near the poles, sunlight strikes Earth at an angle, which spreads the energy over a larger area.

Figure 6 The stunted growth of these trees is a result of abiotic factors.

Elevation If you have climbed or driven up a mountain, you probably noticed that the temperature got cooler as you went higher. A region's elevation, or distance above sea level, affects its temperature. Earth's atmosphere acts as insulation that traps the Sun's heat. At higher elevations, the atmosphere is thinner than it is at lower elevations. Air becomes warmer when sunlight heats molecules in the air. Because there are fewer molecules at higher elevations, air temperatures there tend to be cooler.

At higher elevations, trees are shorter and the ground is rocky, as shown in **Figure 6.** Above the timberline—the elevation beyond which trees do not grow—plant life is limited to low-growing plants. The tops of some mountains are so cold that no plants can survive. Some mountain peaks are covered with snow year-round.

Applying Math   Solve for an Unknown

TEMPERATURE CHANGES You climb a mountain and record the temperature every 1,000 m of elevation. The temperature is 30°C at 304.8 m, 25°C at 609.6 m, 20°C at 914.4 m, 15°C at 1,219.2 m, and 5°C at 1,828.8 m. Make a graph of the data. Use your graph to predict the temperature at an altitude of 2,133.6 m.

Solution

1 *This is what you know:*   The data can be written as ordered pairs (elevation, temperature). The ordered pairs for these data are (304.8, 30), (609.6, 25), (914.4, 20), (1,219.2, 15), (1,828.8, 5).

2 *This is what you want to find:*   Predict the temperature at an elevation of 2,133.6 m.

3 *This is what you need to do:*   Graph the data by plotting elevation on the *x*-axis and temperature on the *y*-axis.

4 *Predict the temperature at 2,133.6 m:*   Extend the graph line to predict the temperature at 2,133.6 m.

Practice Problems

1. Temperatures on another mountain are 33°C at sea level, 31°C at 125 m, 29°C at 250 m, and 26°C at 425 m. Graph the data and predict the temperature at 550 m.

2. Predict what the temperature would be at 375 m.

 **For more practice, visit** booke.msscience.com/ math_practice

Climate

In Fairbanks, Alaska, winter temperatures may be as low as −52°C, and more than a meter of snow might fall in one month. In Key West, Florida, snow never falls and winter temperatures rarely dip below 5°C. These two cities have different climates. **Climate** refers to an area's average weather conditions over time, including temperature, rainfall or other precipitation, and wind.

For the majority of living things, temperature and precipitation are the two most important components of climate. The average temperature and rainfall in an area influence the type of life found there. Suppose a region has an average temperature of 25°C and receives an average of less than 25 cm of rain every year. It is likely to be the home of cactus plants and other desert life. A region with similar temperatures that receives more than 300 cm of rain every year is probably a tropical rain forest.

Wind Heat energy from the Sun not only determines temperature, but also is responsible for the wind. The air is made up of molecules of gas. As the temperature increases, the molecules spread farther apart. As a result, warm air is lighter than cold air. Colder air sinks below warmer air and pushes it upward, as shown in **Figure 7.** These motions create air currents that are called wind.

INTEGRATE
Career

Farmer Changes in weather have a strong influence in crop production. Farmers sometimes adapt by changing planting and harvesting dates, selecting a different crop, or changing water use. In your Science Journal, describe another profession affected by climate.

Science Online

Topic: Weather Data
Visit booke.msscience.com for Web links to information about recent weather data for your area.

Activity In your Science Journal, describe how these weather conditions affect plants or animals that live in your area.

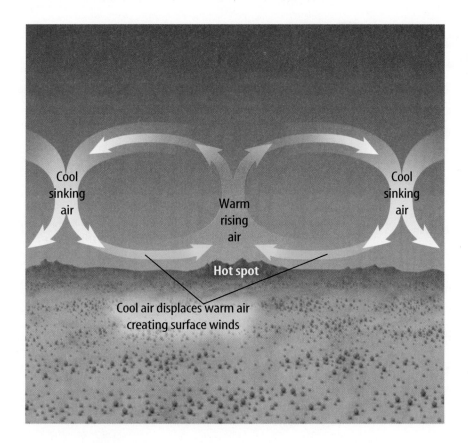

Cool sinking air

Warm rising air

Cool sinking air

Hot spot

Cool air displaces warm air creating surface winds

Figure 7 Winds are created when sunlight heats some portions of Earth's surface more than others. In areas that receive more heat, the air becomes warmer. Cold air sinks beneath the warm air, forcing the warm air upward.

Cold air
loses moisture

Air cools
as it rises

Moist air

Forest

Ocean

Cool, dry air
descends and warms

Desert

Figure 8 In Washington State, the western side of the Cascade Mountains receives an average of 101 cm of rain each year. The eastern side of the Cascades is in a rain shadow that receives only about 25 cm of rain per year.

INTEGRATE
Earth Science

The Rain Shadow Effect The presence of mountains can affect rainfall patterns. As **Figure 8** shows, wind blowing toward one side of a mountain is forced upward by the mountain's shape. As the air nears the top of the mountain, it cools. When air cools, the moisture it contains falls as rain or snow. By the time the cool air crosses over the top of the mountain, it has lost most of its moisture. The other side of the mountain range receives much less precipitation. It is not uncommon to find lush forests on one side of a mountain range and desert on the other side.

section 1 review

Summary

Environmental Factors

- Organisms depend on one another as well as sunlight, air, water, and soil.

Air, Water, and Soil

- Some of the gases in air provide substances to support life.
- Water is a major component of the cells in all organisms.
- Soil supports plant growth.

Sunlight, Temperature, and Climate

- Light energy supports almost all life on Earth.
- Most organisms require temperature between 0°C and 50°C to survive.
- For most organisms, temperature and precipitation are the two most important components of climate.

Self Check

1. **Compare and contrast** biotic factors and abiotic factors in ecosystems.
2. **Explain** why soil is considered an abiotic factor and a biotic factor.
3. **Think Critically** On day 1, you hike in shade under tall trees. On day 2, the trees are shorter and farther apart. On day 3, you see small plants but no trees. On day 4, you see snow. What abiotic factors might contribute to these changes?

Applying Math

4. **Use an Electronic Spreadsheet** Obtain two months of temperature and precipitation data for two cities in your state. Enter the data in a spreadsheet and calculate average daily temperature and precipitation. Use your calculations to compare the two climates.

Humus Farm

Besides abiotic factors, such as rock particles and minerals, soil also contains biotic factors, including bacteria, molds, fungi, worms, insects, and decayed organisms. Crumbly, dark brown soil contains a high percentage of humus that is formed primarily from the decayed remains of plants, animals, and animal droppings. In this lab, you will cultivate your own humus.

Real-World Question

How does humus form?

Goals

- **Observe** the formation of humus.
- **Observe** biotic factors in the soil.
- **Infer** how humus forms naturally.

Materials

widemouthed jar	water
soil	marker
grass clippings or green leaves	metric ruler
	graduated cylinder

Safety Precautions

Wash your hands thoroughly after handling soil, grass clippings, or leaves.

Procedure

1. Copy the data table below into your Science Journal.
2. Place 4 cm of soil in the jar. Pour 30 mL of water into the jar to moisten the soil.
3. Place 2 cm of grass clippings or green leaves on top of the soil in the jar.
4. Use a marker to mark the height of the grass clippings or green leaves in the jar.
5. Put the jar in a sunny place. Every other day, add 30 mL of water to it. In your Science Journal, write a prediction of what you think will happen in your jar.
6. **Observe** your jar every other day for four weeks. Record your observations in your data table.

Conclude and Apply

1. **Describe** what happened during your investigation.
2. **Infer** how molds and bacteria help the process of humus formation.
3. **Infer** how humus forms on forest floors or in grasslands.

Humus Formation

Date	Observations
	Do not write in this book.

Communicating Your Data

Compare your humus farm with those of your classmates. With several classmates, write a recipe for creating the richest humus. Ask your teacher to post your recipe in the classroom. **For more help, refer to the** Science Skill Handbook.

Cycles in Nature

as you read

What You'll Learn

- **Explain** the importance of Earth's water cycle.
- **Diagram** the carbon cycle.
- **Recognize** the role of nitrogen in life on Earth.

Why It's Important

The recycling of matter on Earth demonstrates natural processes.

Review Vocabulary
biosphere: the part of the world in which life can exist

New Vocabulary
- evaporation
- condensation
- water cycle
- nitrogen fixation
- nitrogen cycle
- carbon cycle

The Cycles of Matter

Imagine an aquarium containing water, fish, snails, plants, algae, and bacteria. The tank is sealed so that only light can enter. Food, water, and air cannot be added. Will the organisms in this environment survive? Through photosynthesis, plants and algae produce their own food. They also supply oxygen to the tank. Fish and snails take in oxygen and eat plants and algae. Wastes from fish and snails fertilize plants and algae. Organisms that die are decomposed by the bacteria. The organisms in this closed environment can survive because the materials are recycled. A constant supply of light energy is the only requirement. Earth's biosphere also contains a fixed amount of water, carbon, nitrogen, oxygen, and other materials required for life. These materials cycle through the environment and are reused by different organisms.

The Water Cycle

If you leave a glass of water on a sunny windowsill, the water will evaporate. **Evaporation** takes place when liquid water changes into water vapor, which is a gas, and enters the atmosphere, shown in **Figure 9.** Water evaporates from the surfaces of lakes, streams, puddles, and oceans. Water vapor enters the atmosphere from plant leaves in a process known as transpiration (trans puh RAY shun). Animals release water vapor into the air when they exhale. Water also returns to the environment from animal wastes.

Figure 9 Water vapor is a gas that is present in the atmosphere.

Transpiration

Precipitation

Condensation

Evaporation

Groundwater

Condensation Water vapor that has been released into the atmosphere eventually comes into contact with colder air. The temperature of the water vapor drops. Over time, the water vapor cools enough to change back into liquid water. The process of changing from a gas to a liquid is called **condensation.** Water vapor condenses on particles of dust in the air, forming tiny droplets. At first, the droplets clump together to form clouds. When they become large and heavy enough, they fall to the ground as rain or other precipitation. As the diagram in **Figure 10** shows, the **water cycle** is a model that describes how water moves from the surface of Earth to the atmosphere and back to the surface again.

Figure 10 The water cycle involves evaporation, condensation, and precipitation. Water molecules can follow several pathways through the water cycle.
Identify *as many water cycle pathways as you can from this diagram.*

Water Use Data about the amount of water people take from reservoirs, rivers, and lakes for use in households, businesses, agriculture, and power production is shown in **Table 1.** These actions can reduce the amount of water that evaporates into the atmosphere. They also can influence how much water returns to the atmosphere by limiting the amount of water available to plants and animals.

Table 1 U.S. Estimated Water Use in 1995

Water Use	Millions of Gallons per Day	Percent of Total
Homes and Businesses	41,600	12.2
Industry and Mining	28,000	8.2
Farms and Ranches	139,200	40.9
Electricity Production	131,800	38.7

The Nitrogen Cycle

The element nitrogen is important to all living things. Nitrogen is a necessary ingredient of proteins. Proteins are required for the life processes that take place in the cells of all organisms. Nitrogen is also an essential part of the DNA of all organisms. Although nitrogen is the most plentiful gas in the atmosphere, most organisms cannot use nitrogen directly from the air. Plants need nitrogen that has been combined with other elements to form nitrogen compounds. Through a process called **nitrogen fixation,** some types of soil bacteria can form the nitrogen compounds that plants need. Plants absorb these nitrogen compounds through their roots. Animals obtain the nitrogen they need by eating plants or other animals. When dead organisms decay, the nitrogen in their bodies returns to the soil or to the atmosphere. This transfer of nitrogen from the atmosphere to the soil, to living organisms, and back to the atmosphere is called the **nitrogen cycle,** shown in **Figure 11.**

☑ **Reading Check**   *What is nitrogen fixation?*

Figure 11 During the nitrogen cycle, nitrogen gas from the atmosphere is converted to a soil compound that plants can use. **State** *one source of recycled nitrogen.*

Nitrogen gas is changed into usable compounds by lightning or by nitrogen-fixing bacteria that live on the roots of certain plants.

Plants use nitrogen compounds to build cells.

Animals eat plants. Animal wastes return some nitrogen compounds to the soil.

Animals and plants die and decompose, releasing nitrogen compounds back into the soil.

Figure 12 The swollen nodules on the roots of soybean plants contain colonies of nitrogen-fixing bacteria that help restore nitrogen to the soil. The bacteria depend on the plant for food, while the plant depends on the bacteria to form the nitrogen compounds the plant needs.

Soybeans

Nodules on roots

Nitrogen-fixing bacteria

Stained LM Magnification: 1000×

Soil Nitrogen Human activities can affect the part of the nitrogen cycle that takes place in the soil. If a farmer grows a crop, such as corn or wheat, most of the plant material is taken away when the crop is harvested. The plants are not left in the field to decay and return their nitrogen compounds to the soil. If these nitrogen compounds are not replaced, the soil could become infertile. You might have noticed that adding fertilizer to soil can make plants grow greener, bushier, or taller. Most fertilizers contain the kinds of nitrogen compounds that plants need for growth. Fertilizers can be used to replace soil nitrogen in crop fields, lawns, and gardens. Compost and animal manure also contain nitrogen compounds that plants can use. They also can be added to soil to improve fertility.

Another method farmers use to replace soil nitrogen is to grow nitrogen-fixing crops. Most nitrogen-fixing bacteria live on or in the roots of certain plants. Some plants, such as peas, clover, and beans, including the soybeans shown in **Figure 12,** have roots with swollen nodules that contain nitrogen-fixing bacteria. These bacteria supply nitrogen compounds to the soybean plants and add nitrogen compounds to the soil.

Mini LAB

Comparing Fertilizers

Procedure

1. Examine the three numbers (e.g., 5-10-5) on the **labels of three brands of houseplant fertilizer.** The numbers indicate the percentages of nitrogen, phosphorus, and potassium, respectively, that the product contains.
2. Compare the prices of the three brands of fertilizer.
3. Compare the amount of each brand needed to fertilize a typical houseplant.

Analysis

1. **Identify** the brand with the highest percentage of nitrogen.
2. **Calculate** which brand is the most expensive source of nitrogen. The least expensive.

Figure 13

Carbon—in the form of different kinds of carbon-containing molecules—moves through an endless cycle. The diagram below shows several stages of the carbon cycle. It begins when plants and algae remove carbon from the environment during photosynthesis. This carbon returns to the atmosphere via several carbon-cycle pathways.

A Air contains carbon in the form of carbon dioxide gas. Plants and algae use carbon dioxide to make sugars, which are energy-rich, carbon-containing compounds.

B Organisms break down sugar molecules made by plants and algae to obtain energy for life and growth. Carbon dioxide is released as a waste.

C Burning fossil fuels and wood releases carbon dioxide into the atmosphere.

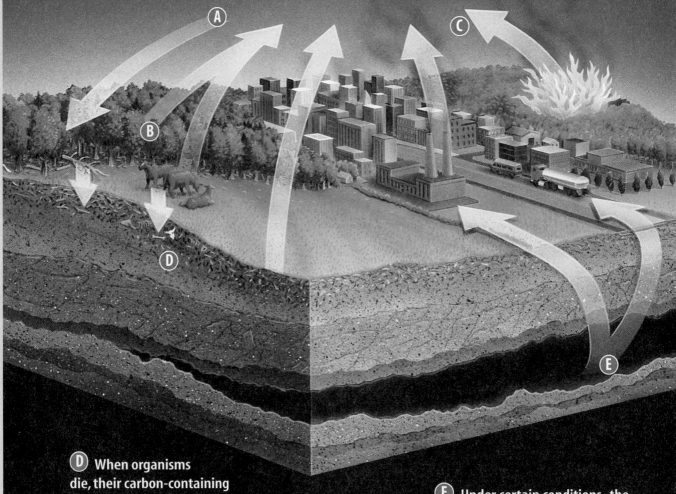

D When organisms die, their carbon-containing molecules become part of the soil. The molecules are broken down by fungi, bacteria, and other decomposers. During this decay process, carbon dioxide is released into the air.

E Under certain conditions, the remains of some dead organisms may gradually be changed into fossil fuels such as coal, gas, and oil. These carbon compounds are energy rich.

The Carbon Cycle

Carbon atoms are found in the molecules that make up living organisms. Carbon is an important part of soil humus, which is formed when dead organisms decay, and it is found in the atmosphere as carbon dioxide gas (CO_2). The **carbon cycle** describes how carbon molecules move between the living and nonliving world, as shown in **Figure 13.**

The carbon cycle begins when producers remove CO_2 from the air during photosynthesis. They use CO_2, water, and sunlight to produce energy-rich sugar molecules. Energy is released from these molecules during respiration—the chemical process that provides energy for cells. Respiration uses oxygen and releases CO_2. Photosynthesis uses CO_2 and releases oxygen. These two processes help recycle carbon on Earth.

✔ **Reading Check** *How does carbon dioxide enter the atmosphere?*

Human activities also release CO_2 into the atmosphere. Fossil fuels such as gasoline, coal, and heating oil are the remains of organisms that lived millions of years ago. These fuels are made of energy-rich, carbon-based molecules. When people burn these fuels, CO_2 is released into the atmosphere as a waste product. People also use wood for construction and for fuel. Trees that are harvested for these purposes no longer remove CO_2 from the atmosphere during photosynthesis. The amount of CO_2 in the atmosphere is increasing. Extra CO_2 could trap more heat from the Sun and cause average temperatures on Earth to rise.

Science Online

Topic: Life Processes
Visit booke.msscience.com for Web links to information about chemical equations that describe photosynthesis and respiration.

Activity Use these equations to explain how respiration is the reverse of photosynthesis.

section 2 review

Summary

The Cycles of Matter
- Earth's biosphere contains a fixed amount of water, carbon, nitrogen, oxygen, and other materials that cycle through the environment.

The Water Cycle
- Water cycles through the environment using several pathways.

The Nitrogen Cycle
- Some types of bacteria can form nitrogen compounds that plants and animals can use.

The Carbon Cycle
- Producers remove CO_2 from the air during photosynthesis and produce O_2.
- Consumers remove O_2 and produce CO_2.

Self Check

1. **Describe** the water cycle.
2. **Infer** how burning fossil fuels might affect the makeup of gases in the atmosphere.
3. **Explain** why plants, animals, and other organisms need nitrogen.
4. **Think Critically** Most chemical fertilizers contain nitrogen, phosphorous, and potassium. If they do not contain carbon, how do plants obtain carbon?

Applying Skills

5. **Identify and Manipulate Variables and Controls** Describe an experiment that would determine whether extra carbon dioxide enhances the growth of tomato plants.

Energy Flow

as you read

What You'll Learn

- **Explain** how organisms produce energy-rich compounds.
- **Describe** how energy flows through ecosystems.
- **Recognize** how much energy is available at different levels in a food chain.

Why It's Important

All living things, including people, need a constant supply of energy.

🔍 Review Vocabulary

energy: the capacity for doing work

New Vocabulary

- chemosynthesis
- food web
- energy pyramid

Converting Energy

All living things are made of matter, and all living things need energy. Matter and energy move through the natural world in different ways. Matter can be recycled over and over again. The recycling of matter requires energy. Energy is not recycled, but it is converted from one form to another. The conversion of energy is important to all life on Earth.

Photosynthesis During photosynthesis, producers convert light energy into the chemical energy in sugar molecules. Some of these sugar molecules are broken down as energy. Others are used to build complex carbohydrate molecules that become part of the producer's body. Fats and proteins also contain stored energy.

Chemosynthesis Not all producers rely on light for energy. During the 1970s, scientists exploring the ocean floor were amazed to find communities teeming with life. These communities were at a depth of almost 3.2 km and living in total darkness. They were found near powerful hydrothermal vents like the one shown in **Figure 14.**

Figure 14 Chemicals in the water that flows from hydrothermal vents provide bacteria with a source of energy. The bacterial producers use this energy to make nutrients through the process of chemosynthesis. Consumers, such as tubeworms, feed on the bacteria.

Hydrothermal Vents A hydrothermal vent is a deep crack in the ocean floor through which the heat of molten magma can escape. The water from hydrothermal vents is extremely hot from contact with molten rock that lies deep in Earth's crust.

Because no sunlight reaches these deep ocean regions, plants or algae cannot grow there. How do the organisms living in this community obtain energy? Scientists learned that the hot water contains nutrients such as sulfur molecules that bacteria use to produce their own food. The production of energy-rich nutrient molecules from chemicals is called **chemosynthesis** (kee moh SIHN thuh sus). Consumers living in the hydrothermal vent communities rely on chemosynthetic bacteria for nutrients and energy. Chemosynthesis and photosynthesis allow producers to make their own energy-rich molecules.

Reading Check *What is chemosynthesis?*

Energy Transfer

Energy can be converted from one form to another. It also can be transferred from one organism to another. Consumers cannot make their own food. Instead, they obtain energy by eating producers or other consumers. The energy stored in the molecules of one organism is transferred to another organism. That organism can oxidize food to release energy that it can use for maintenance and growth or is transformed into heat. At the same time, the matter that makes up those molecules is transferred from one organism to another.

Food Chains A food chain is a way of showing how matter and energy pass from one organism to another. Producers—plants, algae, and other organisms that are capable of photosynthesis or chemosynthesis—are always the first step in a food chain. Animals that consume producers such as herbivores are the second step. Carnivores and omnivores—animals that eat other consumers—are the third and higher steps of food chains. One example of a food chain is shown in **Figure 15.**

INTEGRATE Earth Science

Hydrothermal Vents The first hydrothermal vent community discovered was found along the Galápagos rift zone. A rift zone forms where two plates of Earth's crust are spreading apart. In your Science Journal, describe the energy source that heats the water in the hydrothermal vents of the Galápagos rift zone.

Figure 15 In this food chain, grasses are producers, marmots are herbivores that eat the grasses, and grizzly bears are consumers that eat marmots. The arrows show the direction in which matter and energy flow. **Infer** *what might happen if grizzly bears disappeared from this ecosystem.*

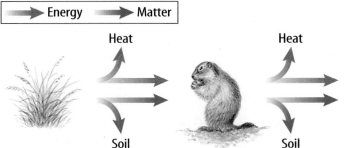

Energy ⟶ Matter ⟶

Heat Heat Heat

Soil Soil Soil

Food Webs A forest community includes many feeding relationships. These relationships can be too complex to show with a food chain. For example, grizzly bears eat many different organisms, including berries, insects, chipmunks, and fish. Berries are eaten by bears, birds, insects, and other animals. A bear carcass might be eaten by wolves, birds, or insects. A **food web** is a model that shows all the possible feeding relationships among the organisms in a community. A food web is made up of many different food chains, as shown in **Figure 16.**

Energy Pyramids

Food chains usually have at least three links, but rarely more than five. This limit exists because the amount of available energy is reduced as you move from one level to the next in a food chain. Imagine a grass plant that absorbs energy from the Sun. The plant uses some of this energy to grow and produce seeds. Some of the energy is stored in the seeds.

Figure 16 Compared to a food chain, a food web provides a more complete model of the feeding relationships in a community.

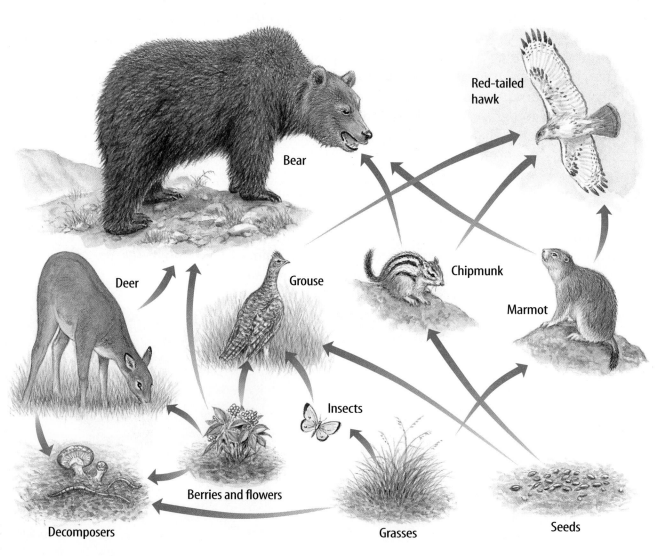

Red-tailed hawk

Bear

Deer

Grouse

Chipmunk

Marmot

Insects

Berries and flowers

Decomposers

Grasses

Seeds

Available Energy When a mouse eats grass seeds, energy stored in the seeds is transferred to the mouse. However, most of the energy the plant absorbed from the Sun was used for the plant's growth. The mouse uses energy from the seed for its own life processes, including respiration, digestion, and growth. Some of this energy was given off as heat. A hawk that eats the mouse obtains even less energy. The amount of available energy is reduced from one feeding level of a food chain to another.

An **energy pyramid,** like the one in **Figure 17,** shows the amount of energy available at each feeding level in an ecosystem. The bottom of the pyramid, which represents all of the producers, is the first feeding level. It is the largest level because it contains the most energy and the largest number of organisms. As you move up the pyramid, the transfer of energy is less efficient and each level becomes smaller. Only about ten percent of the energy available at each feeding level of an energy pyramid is transferred to the next higher level.

Figure 17 An energy pyramid shows that each feeding level has less energy than the one below it. **Describe** what would happen if the hawks and snakes outnumbered the rabbits and mice in this ecosystem.

✔ **Reading Check** *Why does the first feeding level of an energy pyramid contain the most energy?*

section 3 review

Summary

Converting Energy
- Most producers convert light energy into chemical energy.
- Some producers can produce their own food using energy in chemicals such as sulfur.

Energy Transfer
- Producers convert energy into forms that other organisms can use.
- Food chains show how matter and energy pass from one organism to another.

Energy Pyramids
- Energy pyramids show the amount of energy available at each feeding level.
- The amount of available energy decreases from the base to the top of the energy pyramid.

Self Check

1. **Compare and contrast** a food web and an energy pyramid.
2. **Explain** why there is a limit to the number of links in a food chain.
3. **Think Critically** Use your knowledge of food chains and the energy pyramid to explain why the number of mice in a grassland ecosystem is greater than the number of hawks.

Applying Math

4. **Solve One-Step Equations** A forest has 24,055,000 kilocalories (kcals) of producers, 2,515,000 kcals of herbivores, and 235,000 kcals of carnivores. How much energy is lost between producers and herbivores? Between herbivores and carnivores?

Where does the mass of a plant come from?

Real-World Question

An enormous oak tree starts out as a tiny acorn. The acorn sprouts in dark, moist soil. Roots grow down through the soil. Its stem and leaves grow up toward the light and air. Year after year, the tree grows taller, its trunk grows thicker, and its roots grow deeper. It becomes a towering oak that produces thousands of acorns of its own. An oak tree has much more mass than an acorn. Where does this mass come from? The soil? The air? In this activity, you'll find out by conducting an experiment with radish plants. Does all of the matter in a radish plant come from the soil?

Goals

■ **Measure** the mass of soil before and after radish plants have been grown in it.

■ **Measure** the mass of radish plants grown in the soil.

■ **Analyze** the data to determine whether the mass gained by the plants equals the mass lost by the soil.

Materials

8-oz plastic or paper cup
potting soil to fill cup
scale or balance
radish seeds (4)
water
paper towels

Safety Precautions

Procedure

1. Copy the data table into your Science Journal.

2. Fill the cup with dry soil.

3. Find the mass of the cup of soil and record this value in your data table.

4. Moisten the soil in the cup. Plant four radish seeds 2 cm deep in the soil. Space the seeds an equal distance apart. Wash your hands.

5. Add water to keep the soil barely moist as the seeds sprout and grow.

6. When the plants have developed four to six true leaves, usually after two to three weeks, carefully remove the plants from the soil. Gently brush the soil off the roots. Make sure all the soil remains in the cup.

7. Spread the plants out on a paper towel. Place the plants and the cup of soil in a warm area to dry out.

8. When the plants are dry, measure their mass and record this value in your data table. Write this number with a plus sign in the Gain or Loss column.

9. When the soil is dry, find the mass of the cup of soil. Record this value in your data table. Subtract the End mass from the Start mass and record this number with a minus sign in the Gain or Loss column.

Mass of Soil and Radish Plants			
	Start	End	Gain (+) or Loss (−)
Mass of dry soil and cup	Do not write in this book.		
Mass of dried radish plants	0 g		

Analyze Your Data

1. **Calculate** how much mass was gained or lost by the soil. By the radish plants.

2. Did the mass of the plants come completely from the soil? How do you know?

Conclude and Apply

1. In the early 1600s, a Belgian scientist named J. B. van Helmont conducted this experiment with a willow tree. What is the advantage of using radishes instead of a tree?

2. **Predict** where all of the mass gained by the plants came from.

Communicating Your Data

Compare your conclusions with those of other students in your class. **For more help, refer to the** Science Skill Handbook.

Extreme Climates

Did you know...

2,896 cm

... The greatest snowfall in one year

occurred at Mount Baker in Washington State. Approximately 2,896 cm of snow fell during the 1998–99, 12-month snowfall season. That's enough snow to bury an eight-story building.

Applying Math What was the average monthly snowfall at Mount Baker during the 1998–99 snowfall season?

... The hottest climate in the United States

is found in Death Valley, California. In July 1913, Death Valley reached approximately 57°C. As a comparison, a comfortable room temperature is about 20°C.

... The record low temperature

of a frigid −89°C was set in Antarctica in 1983. As a comparison, the temperature of a home freezer is about −15°C.

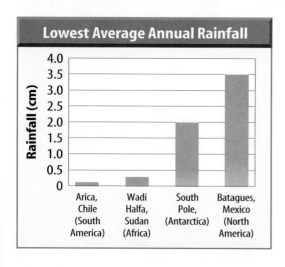

Lowest Average Annual Rainfall

Rainfall (cm)

4.0 3.5 3.0 2.5 2.0 1.5 1.0 0.5 0

Arica, Chile (South America) | Wadi Halfa, Sudan (Africa) | South Pole, (Antarctica) | Batagues, Mexico (North America)

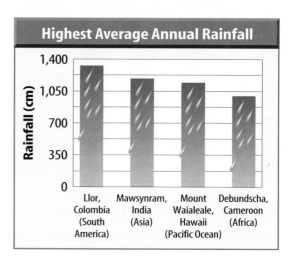

Highest Average Annual Rainfall

Rainfall (cm)

1,400 1,050 700 350 0

Llor, Colombia (South America) | Mawsynram, India (Asia) | Mount Waialeale, Hawaii (Pacific Ocean) | Debundscha, Cameroon (Africa)

Graph It

Visit booke.msscience.com/science_stats **to find the average monthly rainfall in a tropical rain forest. Make a line graph to show how the amount of precipitation changes during the 12 months of the year.**

Reviewing Main Ideas

Section 1   Abiotic Factors

1. Abiotic factors include air, water, soil, sunlight, temperature, and climate.

2. The availability of water and light influences where life exists on Earth.

3. Soil and climate have an important influence on the types of organisms that can survive in different environments.

4. High latitudes and elevations generally have lower average temperatures.

Section 2   Cycles in Nature

1. Matter is limited on Earth and is recycled through the environment.

2. The water cycle involves evaporation, condensation, and precipitation.

3. The carbon cycle involves photosynthesis and respiration.

4. Nitrogen in the form of soil compounds enters plants, which then are consumed by other organisms.

Section 3   Energy Flow

1. Producers make energy-rich molecules through photosynthesis or chemosynthesis.

2. When organisms feed on other organisms, they obtain matter and energy.

3. Matter can be recycled, but energy cannot.

4. Food webs are models of the complex feeding relationships in communities.

5. Available energy decreases as you go to higher feeding levels in an energy pyramid.

Visualizing Main Ideas

This diagram represents photosynthesis in a leaf. Match each letter with one of the following terms: light, carbon dioxide, *or* oxygen.

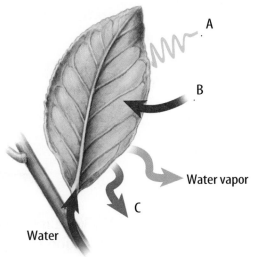

Water vapor

Water

A

B

C

Using Vocabulary

abiotic p. 36	energy pyramid p. 53
atmosphere p. 37	evaporation p. 44
biotic p. 36	food web p. 52
carbon cycle p. 49	nitrogen cycle p. 46
chemosynthesis p. 51	nitrogen fixation p. 46
climate p. 41	soil p. 38
condensation p. 45	water cycle p. 45

Which vocabulary word best corresponds to each of the following events?

1. A liquid changes to a gas.

2. Some types of bacteria form nitrogen compounds in the soil.

3. Decaying plants add nitrogen to the soil.

4. Chemical energy is used to make energy-rich molecules.

5. Decaying plants add carbon to the soil.

6. A gas changes to a liquid.

7. Water flows downhill into a stream. The stream flows into a lake, and water evaporates from the lake.

8. Burning coal and exhaust from automobiles release carbon into the air.

Checking Concepts

Choose the word or phrase that best answers the question.

9. Which of the following is an abiotic factor?
   A) penguins   C) soil bacteria
   B) rain   D) redwood trees

Use the equation below to answer question 10.

$$CO_2 + H_2O \xrightarrow{\text{light energy}} \text{sugar} + O_2$$

10. Which of the following processes is shown in the equation above?
    A) condensation   C) burning
    B) photosynthesis D) respiration

11. Which of the following applies to latitudes farther from the equator?
    A) higher elevations
    B) higher temperatures
    C) higher precipitation levels
    D) lower temperatures

12. Water vapor forming droplets that form clouds directly involves which process?
    A) condensation   C) evaporation
    B) respiration   D) transpiration

13. Which one of the following components of air is least necessary for life on Earth?
    A) argon   C) carbon dioxide
    B) nitrogen   D) oxygen

14. Which group makes up the largest level of an energy pyramid?
    A) herbivores   C) decomposers
    B) producers   D) carnivores

15. Earth receives a constant supply of which of the following items?
    A) light energy   C) nitrogen
    B) carbon   D) water

16. Which of these is an energy source for chemosynthesis?
    A) sunlight   C) sulfur molecules
    B) moonlight   D) carnivores

Use the illustration below to answer question 17.

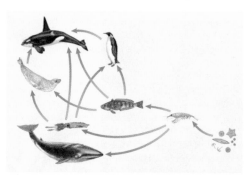

17. What is the illustration above an example of?
    A) food chain   C) energy pyramid
    B) food web   D) carbon cycle

Science Online booke.msscience.com/vocabulary_puzzlemaker

Thinking Critically

18. Draw a Conclusion A country has many starving people. Should they grow vegetables and corn to eat, or should they grow corn to feed cattle so they can eat beef? Explain.

19. Explain why a food web is a better model of energy flow than a food chain.

20. Infer Do bacteria need nitrogen? Why or why not?

21. Describe why it is often easier to walk through an old, mature forest of tall trees than through a young forest of small trees.

22. Explain why giant sequoia trees grow on the west side of California's Inyo Mountains and Death Valley, a desert, is on the east side of the mountains.

23. Concept Map Copy and complete this food web using the following information: *caterpillars and rabbits eat grasses, raccoons eat rabbits and mice, mice eat grass seeds,* and *birds eat caterpillars.*

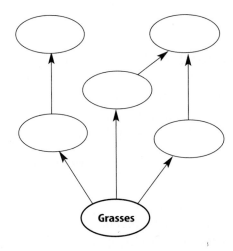

Grasses

24. Form a Hypothesis For each hectare of land, ecologists found 10,000 kcals of producers, 10,000 kcals of herbivores, and 2,000 kcals of carnivores. Suggest a reason why producer and herbivore levels are equal.

25. Recognize Cause and Effect A lake in Kenya has been taken over by a floating weed. How could you determine if nitrogen fertilizer runoff from farms is causing the problem?

Performance Activities

26. Poster Use magazine photographs to make a visual representation of the water cycle.

Applying Math

27. Energy Budget Raymond Lindeman, from the University of Minnesota, was the first person to calculate the total energy budget of an entire community at Cedar Bog Lake in MN. He found the total amount of energy produced by producers was 1,114 kilocalories per meter squared per year. About 20% of the 1,114 kilocalories were used up during respiration. How many kilocalories were used during respiration?

28. Kilocalorie Use Of the 600 kilocalories of producers available to a caterpillar, the caterpillar consumes about 150 kilocalories. About 25% of the 150 kilocalories is used to maintain its life processes and is lost as heat, while 16% cannot be digested. How many kilocalories are lost as heat? What percentage of the 600 kilocalories is available to the next feeding level?

Use the table below to answer question 29.

Mighty Migrators	
Species	**Distance (km)**
Desert locust	4,800
Caribou	800
Green turtle	1,900
Arctic tern	35,000
Gray whale	19,000

29. Make and Use Graphs Climate can cause populations to move from place to place. Make a bar graph of migration distances shown above.

Part 1 Multiple Choice

Record your answers on the answer sheet provided by your teacher or on a sheet of paper.

1. The abiotic factor that provides energy for nearly all life on Earth is
   A. air.   C. water.
   B. sunlight. D. soil.

2. Which of the following is characteristic of places at high elevations?
   A. fertile soil
   B. fewer molecules in the air
   C. tall trees
   D. warm temperatures

Use the diagram below to answer questions 3 and 4.

3. The air at point C is
   A. dry and warm.
   B. dry and cool.
   C. moist and warm.
   D. moist and cool.

4. The air at point A is
   A. dry and warm.
   B. dry and cool.
   C. moist and warm.
   D. moist and cool.

5. What process do plants use to return water vapor to the atmosphere?
   A. transpiration   C. respiration
   B. evaporation   D. condensation

6. Clouds form as a result of what process?
   A. evaporation   C. respiration
   B. transpiration   D. condensation

Use the illustration of the nitrogen cycle below to answer questions 7 and 8.

7. Which of the following items shown in the diagram contribute to the nitrogen cycle by releasing AND absorbing nitrogen?
   A. the decaying organism only
   B. the trees only
   C. the trees and the grazing cows
   D. the lightning and the decaying organism

8. Which of the following items shown in the diagram contribute to the nitrogen cycle by ONLY releasing nitrogen?
   A. the decaying organism only
   B. the trees only
   C. the trees and the grazing cows
   D. the lightning and the decaying organism

9. Where is most of the energy found in an energy pyramid?
   A. at the top level
   B. in the middle levels
   C. at the bottom level
   D. all levels are the same

10. What organisms remove carbon dioxide gas from the air during photosynthesis?
    A. consumers   C. herbivores
    B. producers   D. omnivores

Part 2 | Short Response/Grid In

Record your answers on the answer sheet provided by your teacher or on a sheet of paper.

11. Give two examples of abiotic factors and describe how each one is important to biotic factors.

Use the table below to answer questions 12 and 13.

U.S. Estimated Water Use in 1995		
Water Use	Millions of Gallons per Day	Percent of Total
Homes and Businesses	41,600	12.2
Industry and Mining	28,000	8.2
Farms and Ranches	139,200	40.9
Electricity Production	131,800	38.7

12. According to the table above, what accounted for the highest water use in the U.S. in 1995?

13. What percentage of the total amount of water use results from electricity production and homes and business combined?

14. Where are nitrogen-fixing bacteria found?

15. Describe two ways that carbon is released into the atmosphere.

16. How are organisms near hydrothermal vents deep in the ocean able to survive?

17. Use a diagram to represent the transfer of energy among these organisms: a weasel, a rabbit, grasses, and a coyote.

Test-Taking Tip

Answer All Questions Never leave any answer blank.

Part 3 | Open Ended

Record your answers on a sheet of paper.

18. Explain how a decrease in the amount of sunlight would affect producers that use photosynthesis, and producers that use chemosynthesis.

19. Describe how wind and wind currents are produced.

20. Use the water cycle to explain why beads of water form on the outside of a glass of iced water on a hot day.

21. Draw a flowchart that shows how soy beans, deer, and nitrogen-fixing bacteria help cycle nitrogen from the atmosphere, to the soil, to living organisms, and back to the atmosphere.

Use the diagram below to answer questions 22 and 23.

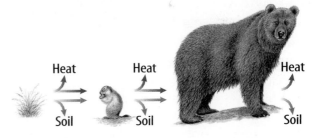

22. What term is used for the diagram above? Explain how the diagram represents energy transfer.

23. Explain how the grass and bear populations would be affected if the marmot population suddenly declined.

24. Compare and contrast an energy pyramid and a food web.

25. What happens to the energy in organisms at the top of an energy pyramid when they die?

Ecosystems

The BIG Idea

Earth has many diverse ecosystems on land and in water.

SECTION 1
How Ecosystems Change
Main Idea Ecosystems gradually change over time.

SECTION 2
Biomes
Main Idea Land on Earth is divided into large geographic areas that have similar climates and ecosystems.

SECTION 3
Acquatic Ecosystems
Main Idea Both Earth's salt water and freshwater are divided into a variety of ecosystems.

The Benefits of Wildfires

Ecosystems are places where organisms, including humans, interact with each other and with their physical environment. In some ecosystems, wildfires are an essential part of the physical environment. Organisms in these ecosystems are well adapted to the changes that fire brings, and can benefit from wildfires.

Science Journal What traits might plants on this burning Montana hillside have that enable them to survive?

Start-Up Activities

What environment do houseplants need?

The plants growing in your classroom or home may not look like the same types of plants that you find growing outside. Many indoor plants don't grow well outside in most North American climates. Do the lab below to determine what type of environment most houseplants thrive in.

1. Examine a healthy houseplant in your classroom or home.

2. Describe the environmental conditions found in your classroom or home. For example, is the air humid or dry? Is the room warm or cool? Does the temperature stay about the same, or change during the day?

3. Using observations from step 1 and descriptions from step 2, hypothesize about the natural environment of the plants in your classroom or home.

4. **Think Critically** In your Science Journal, record the observations that led to your hypothesis. How would you design an experiment to test your hypothesis?

Primary and Secondary Succession Make the following Foldable to help you illustrate the main ideas about succession.

STEP 1 Fold a vertical sheet of paper in half from top to bottom.

STEP 2 Fold in half from side to side with the fold at the top.

STEP 3 Unfold the paper once. **Cut** only the fold of the top flap to make two tabs.

STEP 4 Turn the paper vertically and **label** on the front tabs as shown.

primary succession
↓
climax community
secondary succession
↓
climax community

Illustrate and Label As you read the chapter, define terms and collect information under the appropriate tabs.

Science Online
Preview this chapter's content and activities at
booke.msscience.com

Get Ready to Read

Take Notes

① Learn It! The best way for you to remember information is to write it down, or take notes. Good note-taking is useful for studying and research. When you are taking notes, it is helpful to
- phrase the information in your own words;
- restate ideas in short, memorable phrases;
- stay focused on main ideas and only the most important supporting details.

② Practice It! Make note-taking easier by using a chart to help you organize information clearly. Write the main ideas in the left column. Then write at least three supporting details in the right column. Read the text from Section 3 of this chapter under the heading *Water Pollution*, page 89. Then take notes using a chart, such as the one below.

Main Idea	Supporting Details
	1. 2. 3. 4. 5.
	1. 2. 3. 4. 5.

③ Apply It! As you read this chapter, make a chart of the main ideas. Next to each main idea, list at least two supporting details.

Reading Tip

Read one or two paragraphs first and take notes after you read. You are likely to take down too much information if you take notes as you read.

Target Your Reading

Use this to focus on the main ideas as you read the chapter.

1 **Before you read** the chapter, respond to the statements below on your worksheet or on a numbered sheet of paper.

- Write an **A** if you **agree** with the statement.
- Write a **D** if you **disagree** with the statement.

2 **After you read** the chapter, look back to this page to see if you've changed your mind about any of the statements.

- If any of your answers changed, explain why.
- Change any false statements into true statements.
- Use your revised statements as a study guide.

Science Online

Print out a worksheet of this page at booke.msscience.com

Before You Read A or D		Statement	After You Read A or D
	1	Gradual changes to the types of species in an ecosystem always follow the same pattern.	
	2	A variety of organisms can live on bare rock.	
	3	Plant communities are always changing and unstable.	
	4	Deserts in different parts of the world do not have any similarities with one another.	
	5	Rainforests are all located near the equator.	
	6	Most of the soil in the tundra is frozen year round.	
	7	Aquatic ecosystems are either freshwater or saltwater.	
	8	Coral reefs are durable ecosystems that adapt quickly to stress.	
	9	The amount of sunlight available in ocean and lake waters affects the number of organisms found there.	

How Ecosystems Change

What **You'll Learn**

- **Explain** how ecosystems change over time.
- **Describe** how new communities begin in areas without life.
- **Compare** pioneer species and climax communities.

Why **It's Important**

Understanding ecosystems and your role in them can help you manage your impact on them and predict the changes that may happen in the future.

🔎 **Review Vocabulary**

ecosystem: community of living organisms interacting with each other and their physical environment

New Vocabulary

- • succession
- • pioneer species
- • climax community

Ecological Succession

What would happen if the lawn at your home were never cut? The grass would get longer, as in **Figure 1,** and soon it would look like a meadow. Later, larger plants would grow from seeds brought to the area by animals or wind. Then, trees might sprout. In fact, in 20 years or less you wouldn't be able to tell that the land was once a mowed lawn. An ecologist can tell you what type of ecosystem your lawn would become. If it would become a forest, they can tell you how long it would take and predict the type of trees that would grow there. **Succession** refers to the normal, gradual changes that occur in the types of species that live in an area. Succession occurs differently in different places around the world.

Primary Succession As lava flows from the mouth of a volcano, it is so hot that it destroys everything in its path. When it cools, lava forms new land composed of rock. It is hard to imagine that this land eventually could become a forest or grassland someday.

The process of succession that begins in a place previously without plants is called primary succession. It starts with the arrival of living things such as lichens (LI kunz). These living things, called **pioneer species,** are the first to inhabit an area. They survive drought, extreme heat and cold, and other harsh conditions and often start the soil-building process.

Figure 1 Open areas that are not maintained will become overgrown with grasses and shrubs as succession proceeds.

Figure 2 Lichens, like these in Colorado, are fragile and take many years to grow. They often cling to bare rock where many other organisms can't survive. **Describe** *how lichens form soil.*

New Soil During primary succession, shown in **Figure 2,** soil begins to form as lichens and the forces of weather and erosion help break down rocks into smaller pieces. When lichens die, they decay, adding small amounts of organic matter to the rock. Plants such as mosses and ferns can grow in this new soil. Eventually, these plants die, adding more organic material. The soil layer thickens, and grasses, wildflowers, and other plants begin to take over. When these plants die, they add more nutrients to the soil. This buildup is enough to support the growth of shrubs and trees. All the while, insects, small birds, and mammals have begun to move in. What was once bare rock now supports a variety of life.

Secondary Succession What happens when a fire, such as the one in **Figure 3,** disturbs a forest or when a building is torn down in a city? After a forest fire, not much seems to be left except dead trees and ash-covered soil. After the rubble of a building is removed, all that remains is bare soil. However, these places do not remain lifeless for long. The soil already contains the seeds of weeds, grasses, and trees. More seeds are carried to the area by wind and birds. Other wildlife may move in. Succession that begins in a place that already has soil and was once the home of living organisms is called secondary succession. Because soil already is present, secondary succession occurs faster and has different pioneer species than primary succession does.

 *Which type of succession usually starts without soil?*

Topic: Eutrophication
Visit for booke.msscience.com
Web links to information about eutrophication (yoo truh fih KAY shun)—secondary succession in an aquatic ecosystem.

Activity Using the information that you find, illustrate or describe in your Science Journal this process for a small freshwater lake.

Figure 3

In the summer of 1988, wind-driven flames like those shown in the background photo swept through Yellowstone National Park, scorching nearly a million acres. The Yellowstone fire was one of the largest forest fires in United States history. The images on this page show secondary succession—the process of ecological regeneration—triggered by the fire.

▶ After the fire, burned timber and blackened soil seemed to be all that remained. However, the fire didn't destroy the seeds that were protected under the soil.

◀ Within weeks, grasses and other plants were beginning to grow in the burned areas. Ecological succession was underway.

▶ Many burned areas in the park opened new plots for stands of trees. This picture shows young lodgepole pines in August 1999. The forest habitat of America's oldest national park is being restored gradually through secondary succession.

Figure 4 This beech-maple forest is an example of a climax community.

Climax Communities A community of plants that is relatively stable and undisturbed and has reached a stage of succession is called a **climax community.** The beech-maple forest shown in **Figure 4** is an example of a community that has reached the end of succession. New trees grow when larger, older trees die. The individual trees change, but the species remain stable. There are fewer changes of species in a climax community over time, as long as the community isn't disturbed by wildfire, avalanche, or human activities.

Primary succession begins in areas with no previous vegetation. It can take hundreds or even thousands of years to develop into a climax community. Secondary succession is usually a shorter process, but it still can take a century or more.

section 1 review

Summary

Ecological Succession

- Succession is the natural, gradual changes over time of species in a community.
- Primary succession occurs in areas that previously were without soil or plants.
- Secondary succession occurs in areas where soil has been disturbed.
- Climax communities have reached an end stage of succession and are stable.
- Climax communities have less diversity than communities in mid-succession.

Self Check

1. **Compare** primary and secondary succession.
2. **Describe** adaptations of pioneer species.
3. **Infer** the kind of succession that will take place on an abandoned, unpaved country road.
4. **Think Critically** Show the sequence of events in primary succession. Include the term *climax community*.

Applying Math

5. **Solve One-Step Equations** A tombstone etched with 1802 as the date of death has a lichen on it that is 6 cm in diameter. If the lichen began growing in 1802, calculate its average yearly rate of growth.

Biomes

as you read

What You'll Learn

- **Explain** how climate influences land environments.
- **Identify** seven biomes of Earth.
- **Describe** the adaptations of organisms found in each biome.

Why It's Important

Resources that you need to survive are found in a variety of biomes.

Review Vocabulary

climate: the average weather conditions of an area over many years

New Vocabulary

- biome
- tundra
- taiga
- temperate deciduous forest
- temperate rain forest
- tropical rain forest
- desert
- grassland

Factors That Affect Biomes

Does a desert in Arizona have anything in common with a desert in Africa? Both have heat, little rain, poor soil, water-conserving plants with thorns, and lizards. Even widely separated regions of the world can have similar biomes because they have similar climates. Climate is the average weather pattern in an area over a long period of time. The two most important climatic factors that affect life in an area are temperature and precipitation.

Major Biomes

Large geographic areas that have similar climates and ecosystems are called **biomes** (BI ohmz). Seven common types of land biomes are mapped in **Figure 5.** Areas with similar climates produce similar climax communities. Tropical rain forests are climax communities found near the equator, where temperatures are warm and rainfall is plentiful. Coniferous forests grow where winter temperatures are cold and rainfall is moderate.

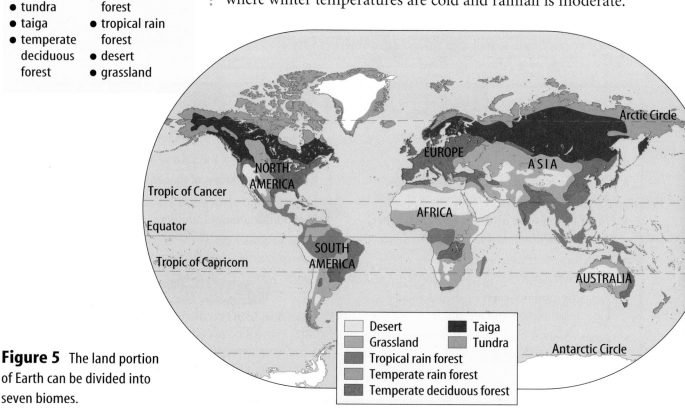

Figure 5 The land portion of Earth can be divided into seven biomes.

Desert	Taiga
Grassland	Tundra
Tropical rain forest	
Temperate rain forest	
Temperate deciduous forest	

Tundra At latitudes just south of the north pole or at high elevations, a biome can be found that receives little precipitation but is covered with ice most of the year. The **tundra** is a cold, dry, treeless region, sometimes called a cold desert. Precipitation averages less than 25 cm per year. Winters in the Arctic can be six to nine months long. For some of these months, the Sun never appears above the horizon and it is dark 24 hours a day. The average daily temperature is about −12°C. For a few days during the short, cold summer, the Sun is always visible. Only the top portion of soil thaws in the summer. Below the thawed surface is a layer of permanently frozen soil called permafrost, shown in **Figure 6.** Alpine tundra, found above the treeline on high mountains, have similar climates. Tundra soil has few nutrients because the cold temperatures slow the process of decomposition.

Figure 6 This permafrost in Alaska is covered by soil that freezes in the winter and thaws in the summer.
Infer *what types of problems this might cause for people living in this area.*

Tundra Life Tundra plants are adapted to drought and cold. They include mosses, grasses, and small shrubs, as seen in **Figure 7.** Many lichens grow on the tundra. During the summer, mosquitoes, blackflies, and other biting insects fill the air. Migratory birds such as ducks, geese, shorebirds, and songbirds nest on the Arctic tundra during the summer. Other inhabitants include hawks, snowy owls, and willow grouse. Mice, voles, lemmings, arctic hares, caribou, reindeer, and musk oxen also are found there.

People are concerned about overgrazing by animals on the tundra. Fences, roads, and pipelines have disrupted the migratory routes of some animals and forced them to stay in a limited area. Because the growing season is so short, plants and other vegetation can take decades to recover from damage.

Figure 7 Lichens, mosses, grasses, and small shrubs thrive on the tundra. Ptarmigan also live on the tundra. In winter, their feathers turn white. Extra feathers on their feet keep them warm and prevent them from sinking into the snow.

Tundra

Ptarmigan

Taiga South of the tundra—between latitudes 50°N and 60°N and stretching across North America, northern Europe, and Asia—is the world's largest biome. The **taiga** (TI guh), shown in **Figure 8,** is a cold, forest region dominated by cone-bearing evergreen trees. Although the winter is long and cold, the taiga is warmer and wetter than the tundra. Precipitation is mostly snow and averages 35 cm to 100 cm each year.

Most soils of the taiga thaw completely during the summer, making it possible for trees to grow. However, permafrost is present in the extreme northern regions of the taiga. The forests of the taiga might be so dense that little sunlight penetrates the trees to reach the forest floor. However, some lichens and mosses do grow on the forest floor. Moose, lynx, shrews, bears, and foxes are some of the animals that live in the taiga.

Figure 8 The taiga is dominated by cone-bearing trees. The lynx, a mammal adapted to life in the taiga, has broad, heavily furred feet that act like snowshoes to prevent it from sinking in the snow.
Infer *why "snowshoe feet" are important for a lynx.*

Temperate Deciduous Forest Temperate regions usually have four distinct seasons each year. Annual precipitation ranges from about 75 cm to 150 cm and is distributed throughout the year. Temperatures range from below freezing during the winter to 30°C or more during the warmest days of summer.

Figure 9 White-tailed deer are one of many species that you can find in a deciduous forest. In autumn, the leaves on deciduous trees change color and fall to the ground.

Temperate Forest Life Many evergreen trees grow in the temperate regions of the world. However, most of the temperate forests in Europe and North America are dominated by climax communities of deciduous trees, which lose their leaves every autumn. These forests, like the one in **Figure 9,** are called **temperate deciduous forests.** In the United States, most of them are located east of the Mississippi River.

When European settlers first came to America, they cut trees to create farmland and to supply wood. As forests were cut, organisms lost their habitats. When agriculture shifted from the eastern to the midwestern and western states, secondary succession began, and trees eventually returned to some areas. Now, nearly as many trees grow in the New England states as did before the American Revolutionary War. Many trees are located in smaller patches. Yet, the recovery of large forests such as those in the Adirondack Mountains in New York State shows the result of secondary succession.

Temperate Rain Forest New Zealand, southern Chile, and the Pacific Northwest of the United States are some of the places where **temperate rain forests,** shown in **Figure 10,** are found. The average temperature of a temperate rain forest ranges from 9°C to 12°C. Precipitation ranges from 200 cm to 400 cm per year.

Trees with needlelike leaves dominate these forests, including the Douglas fir, western red cedar, and spruce. Many grow to great heights. Animals of the temperate rain forest include the black bear, cougar, bobcat, northern spotted owl, and marbled murrelet. Many species of amphibians also inhabit the temperate rain forest, including salamanders.

The logging industry in the Northwest provides jobs for many people. However, it also removes large parts of the temperate rain forest and destroys the habitat of many organisms. Many logging companies now are required to replant trees to replace the ones they cut down. Also, some rain forest areas are protected as national parks and forests.

Figure 10 In the Olympic rain forest in Washington State, mosses and lichens blanket the ground and hang from the trees. Wet areas are perfect habitats for amphibians like the Pacific giant salamander above.

Figure 11 Tropical rain forests are lush environments that contain such a large variety of species that many have not been discovered.

Mini LAB

Modeling Rain Forest Leaves

Procedure

1. Draw an oval leaf about 10 cm long on a piece of **poster board.** Cut it out.
2. Draw a second leaf the same size but make one end pointed. This is called a drip tip. Cut this leaf out.
3. Hold your hands palm-side up over a **sink** and have someone lay a leaf on each one. Point the drip tip away from you. Tilt your hands down but do not allow the leaves to fall off.
4. Have someone gently spray water on the leaves and observe what happens.

Analysis

1. From which leaf does water drain faster?
2. Infer why it is an advantage for a leaf to get rid of water quickly in a rain forest.

Try at Home

Tropical Rain Forest Warm temperatures, wet weather, and lush plant growth are found in **tropical rain forests.** These forests are warm because they are near the equator. The average temperature, about 25°C, doesn't vary much between night and day. Most tropical rain forests receive at least 200 cm of rain annually. Some receive as much as 600 cm of rain each year.

Tropical rain forests, like the one in **Figure 11,** are home to an astonishing variety of organisms. They are one of the most biologically diverse places in the world. For example, one tree in a South American rain forest might contain more species of ants than exist in all of the British Isles.

Tropical Rain Forest Life Different animals and plants live in different parts of the rain forest. Scientists divide the rain forest into zones based on the types of plants and animals that live there, just as a library separates books about different topics onto separate shelves. The zones include: forest floor, understory, canopy, and emergents, as shown in **Figure 12.** These zones often blend together, but their existence provide different habitats for many diverse organisms to live in the tropical rain forest.

✔ Reading Check *What are the four zones of a tropical rain forest?*

Although tropical rain forests support a huge variety of organisms, the soil of the rain forest contains few nutrients. Over the years, nutrients have been washed out of the soil by rain. On the forest floor, decomposers immediately break down organic matter, making nutrients available to the plants again.

Human Impact Farmers that live in tropical areas clear the land to farm and to sell the valuable wood. After a few years, the crops use up the nutrients in the soil, and the farmers must clear more land. As a result, tropical rain forest habitats are being destroyed. Through education, people are realizing the value and potential value of preserving the species of the rain forest. In some areas, logging is prohibited. In other areas, farmers are taught new methods of farming so they do not have to clear rain forest lands continually.

Figure 12 Tropical rain forests contain abundant and diverse organisms.

Emergents These giant trees are much higher than the average canopy tree. Birds, such as the macaw, and insects are found here.

Canopy The canopy includes the upper parts of the trees. It's full of life—insects, birds, reptiles, and mammals.

Understory This dark, cool environment is under the canopy leaves but above the ground. Many insects, reptiles, and amphibians live in the understory.

Forest Floor The forest floor is home to many insects and the largest mammals in the rain forest generally live here.

INTEGRATE Earth Science

Desertification When vegetation is removed from soil in areas that receive little rain, the dry, unprotected surface can be blown away. If the soil remains bare, a desert might form. This process is called desertification. Look on a biome map and hypothesize about which areas of the United States are most likely to become deserts.

Desert The driest biome on Earth is the **desert.** Deserts receive less than 25 cm of rain each year and support little plant life. Some desert areas receive no rain for years. When rain does come, it quickly drains away. Any water that remains on the ground evaporates rapidly.

Most deserts, like the one in **Figure 13,** are covered with a thin, sandy, or gravelly soil that contains little organic matter. Due to the lack of water, desert plants are spaced far apart and much of the ground is bare. Barren, windblown sand dunes are characteristics of the driest deserts.

✔ **Reading Check** *Why is much of a desert bare ground?*

Desert Life Desert plants are adapted for survival in the extreme dryness and hot and cold temperatures of this biome. Most desert plants are able to store water. Cactus plants are probably the most familiar desert plants of the western hemisphere. Desert animals also have adaptations that help them survive the extreme conditions. Some, like the kangaroo rat, never need to drink water. They get all the moisture they need from the breakdown of food during digestion. Most animals are active only during the night, late afternoon, or early morning when temperatures are less extreme. Few large animals are found in the desert.

In order to provide water for desert cities, rivers and streams have been diverted. When this happens, wildlife tends to move closer to cities in their search for food and water. Education about desert environments has led to an awareness of the impact of human activities. As a result, large areas of desert have been set aside as national parks and wilderness areas to protect desert habitats.

Figure 13 Desert plants, like these in the Sonoran Desert, are adapted for survival in the extreme conditions of the desert biome. The giant hairy scorpion found in some deserts has a venomous sting.

Grasslands Temperate and tropical regions that receive between 25 cm and 75 cm of precipitation each year and are dominated by climax communities of grasses are called **grasslands.** Most grasslands have a dry season, when little or no rain falls. This lack of moisture prevents the development of forests. Grasslands are found in many places around the world, and they have a variety of names. The prairies and plains of North America, the steppes of Asia, the savannas of Africa shown in **Figure 14,** and the pampas of South America are types of grasslands.

Grasslands Life The most noticeable animals in grassland biomes are usually mammals that graze on the stems, leaves, and seeds of grass plants. Kangaroos graze in the grasslands of Australia. In Africa, communities of animals such as wildebeests, impalas, and zebras thrive in the savannas.

Grasslands are perfect for growing many crops such as wheat, rye, oats, barley, and corn. Grasslands also are used to raise cattle and sheep. However, overgrazing can result in the death of grasses and the loss of valuable topsoil from erosion. Most farmers and ranchers take precautions to prevent the loss of valuable habitats and soil.

Figure 14 Animals such as zebras and wildebeests are adapted to life on the savannas in Africa.

section 2 review

Summary

Major Biomes

- Tundra, sometimes called a cold desert, can be divided into two types: arctic and alpine.

- Taiga is the world's largest biome. It is a cold forest region with long winters.

- Temperate regions have either a deciduous forest biome or a rain forest biome.

- Tropical rain forests are one of the most biologically diverse biomes.

- Humans have a huge impact on tropical rain forests.

- The driest biome is the desert. Desert organisms are adapted for extreme dryness and temperatures.

- Grasslands provide food for wildlife, livestock, and humans.

Self Check

1. **Determine** which two biomes are the driest.

2. **Compare and contrast** tundra organisms and desert organisms.

3. **Identify** the biggest climatic difference between a temperate rain forest and a tropical rain forest.

4. **Explain** why the soil of tropical rain forests make poor farmland.

5. **Think Critically** If you climb a mountain in Arizona, you might reach an area where the trees resemble the taiga trees in northern Canada. Why would a taiga forest exist in Arizona?

Applying Skills

6. **Record Observations** Animals have adaptations that help them survive in their environments. Make a list of animals that live in your area, and record the physical or behavioral adaptations that help them survive.

Studying a Land Ecosystem

An ecological study includes observation and analysis of organisms and the physical features of the environment.

Real-World Question

How do you study an ecosystem?

Goals
■ **Observe** biotic factors and abiotic factors of an ecosystem.
■ **Analyze** the relationships among organisms and their environments.

Materials

graph paper   field guides
binoculars   notebook
thermometer   compass
pencil   tape measure
magnifying lens

Safety Precautions

Procedure

1. Choose a portion of an ecosystem to study. You might choose a decaying log, a pond, a garden, or even a crack in the sidewalk.

2. Determine the boundaries of your study area.

3. Using a tape measure and graph paper, make a map of your area. Determine north.

4. **Record** your observations in a table similar to the one shown on this page.

5. **Observe** the organisms in your study area. Use field guides to identify them. Use a magnifying lens to study small organisms and binoculars to study animals you can't get near. Look for evidence (such as tracks or feathers) of organisms you do not see.

6. Measure and record the air temperature in your study area.

7. Visit your study area many times and at different times of day for one week. At each visit, make the same measurements and record all observations. Note how the living and nonliving parts of the ecosystem interact.

Environmental Observations		
Date		
Time of day		
Temperature		
Organisms observed	Do not write in this book.	
Comments		

Conclude and Apply

1. **Predict** what might happen if one or more abiotic factors were changed suddenly.

2. **Infer** what might happen if one or more populations of plants or animals were removed from the area.

3. **Form a hypothesis** to explain how a new population of organisms might affect your ecosystem.

Communicating Your Data

Make a classroom display of all data recorded. **For more help, refer to the Science Skill Handbook.**

Aquatic Ecosystems

Freshwater Ecosystems

In a land environment, temperature and precipitation are the most important factors that determine which species can survive. In aquatic environments, water temperature, the amount of sunlight present, and the amounts of dissolved oxygen and salt in the water are important. Earth's freshwater ecosystems include flowing water such as rivers and streams and standing water such as lakes, ponds, and wetlands.

Rivers and Streams Flowing freshwater environments vary from small, gurgling brooks to large, slow-moving rivers. Currents can quickly wash loose particles downstream, leaving a rocky or gravelly bottom. As the water tumbles and splashes, as shown in **Figure 15,** air from the atmosphere mixes in. Naturally fast-flowing streams usually have clearer water and higher oxygen content than slow-flowing streams.

Most nutrients that support life in flowing-water ecosystems are washed into the water from land. In areas where the water movement slows, such as in the pools of streams or in large rivers, debris settles to the bottom. These environments tend to have higher nutrient levels and more plant growth. They contain organisms that are not as well adapted to swiftly flowing water, such as freshwater mussels, minnows, and leeches.

as you read

What **You'll Learn**
- **Compare** flowing freshwater and standing freshwater ecosystems.
- **Identify** and describe important saltwater ecosystems.
- **Identify** problems that affect aquatic ecosystems.

Why **It's Important**
All of the life processes in your body depend on water.

Review Vocabulary
aquatic: growing or living in water

New Vocabulary
- wetland   • intertidal zone
- coral reef   • estuary

Figure 15 Streams like this one are high in oxygen because of the swift, tumbling water.
Determine *where most nutrients in streams come from.*

Modeling Freshwater Environments

Procedure

1. Obtain a sample of **pond sediment or debris, plants, water, and organisms** from your teacher.
2. Cover the bottom of a **clear-plastic container** with about 2 cm of the debris.
3. Add one or two plants to the container.
4. Carefully pour pond water into the container until it is about two-thirds full.
5. Use a **net** to add several organisms to the water. Seal the container.
6. Using a **magnifying lens,** observe as many organisms as possible. Record your observations. Return your sample to its original habitat.

Analysis
Write a short paragraph describing the organisms in your sample. How did the organisms interact with each other?

Human Impact People use rivers and streams for many activities. Once regarded as a free place to dump sewage and other pollutants, many people now recognize the damage this causes. Treating sewage and restricting pollutants have led to an improvement in the water quality in some rivers.

Lakes and Ponds When a low place in the land fills with rainwater, snowmelt, or water from an overflowing stream, a lake or pond might form. Pond or lake water hardly moves. It contains more plants than flowing-water environments contain.

Lakes, such as the one shown in **Figure 16,** are larger and deeper than ponds. They have more open water because most plant growth is limited to shallow areas along the shoreline. In fact, organisms found in the warm, sunlit waters of the shorelines often are similar to those found in ponds. If you were to dive to the bottom, you would discover few, if any, plants or algae growing. Colder temperatures and lower light levels limit the types of organisms that can live in deep lake waters. Floating in the warm, sunlit waters near the surface of freshwater lakes and ponds are microscopic algae, plants, and other organisms known as plankton.

A pond is a small, shallow body of water. Because ponds are shallow, they are filled with animal and plant life. Sunlight usually penetrates to the bottom. The warm, sunlit water promotes the growth of plants and algae. In fact, many ponds are filled almost completely with plant material, so the only clear, open water is at the center. Because of the lush growth in pond environments, they tend to be high in nutrients.

Figure 16 Ponds contain more vegetation than lakes contain. The population of organisms in the shallow water of lakes is high. Fewer types of organisms live in the deeper water.

Water Pollution Human activities can harm freshwater environments. Fertilizer-filled runoff from farms and lawns, as well as sewage dumped into the water, can lead to excessive growth of algae and plants in lakes and ponds. The growth and decay of these organisms reduces the oxygen level in the water, which makes it difficult for some organisms to survive. To prevent problems, sewage is treated before it is released. People also are being educated about problems associated with polluting lakes and ponds. Fines and penalties are issued to people caught polluting waterways. These controls have led to the recovery of many freshwater ecosystems.

Wetlands As the name suggests, **wetlands,** shown in **Figure 17,** are regions that are wet for all or most of a year. They are found in regions that lie between landmasses and water. Other names for wetlands include swamps, bogs, and fens. Some people refer to wetlands as biological supermarkets. They are fertile ecosystems, but only plants that are adapted to water-logged soil survive there. Wetland animals include beavers, muskrats, alligators, and the endangered bog turtle. Many migratory bird populations use wetlands as breeding grounds.

Reading Check *Where are wetlands found?*

Wetlands once were considered to be useless, disease-ridden places. Many were drained and destroyed to make roads, farmland, shopping centers, and housing developments. Only recently have people begun to understand the importance of wetlands. Products that come from wetlands, including fish, shellfish, cranberries, and plants, are valuable resources. Now many developers are restoring wetlands, and in most states access to land through wetlands is prohibited.

Figure 17 Life in the Florida Everglades was threatened due to pollution, drought, and draining of the water. Conservation efforts are being made in an attempt to preserve this ecosystem.

Environmental Author Rachel Carson (1907–1964) was a scientist that turned her knowledge and love of the environment into articles and books. After 15 years as an editor for the U.S. Fish and Wildlife Service, she resigned and devoted her time to writing. She probably is known best for her book *Silent Spring,* in which she warned about the long-term effects of the misuse of pesticides. In your Science Journal, compile a list of other authors who write about environmental issues.

Saltwater Ecosystems

About 95 percent of the water on the surface of Earth contains high concentrations of various salts. The amount of dissolved salts in water is called salinity. The average ocean salinity is about 35 g of salts per 1,000 g of water. Saltwater ecosystems include oceans, seas, a few inland lakes such as the Great Salt Lake in Utah, coastal inlets, and estuaries.

Applying Math Convert Units

TEMPERATURE Organisms that live around hydrothermal vents in the ocean deal with temperatures that range from 1.7°C to 371°C. You have probably seen temperatures measured in degrees Celsius (°C) and degrees Fahrenheit (°F). Which one are you familiar with? If you know the temperature in one system, you can convert it to the other.

You have a Fahrenheit thermometer and measure the water temperature of a pond at 59°F. What is that temperature in degrees Celsius?

Solution

1 *This is what you know:* water temperature in degrees Fahrenheit = 59°F

2 *This is what you need to find out:* The water temperature in degrees Celsius.

3 *This is the procedure you need to use:*
- Solve the equation for degrees Celsius:
  $$(°C \times 1.8) + 32 = °F$$
  $$°C = (°F - 32)/1.8$$
- Substitute the known value:
  $$°C = (59°F - 32)/1.8 = 15°C$$
- Water temperature that is 59°F is 15°C.

4 *Check your answer:* Substitute the Celsius temperature back into the original equation. You should get 59.

Practice Problems

1. The thermometer outside your classroom reads 78°F. What is the temperature in degrees Celsius?

2. If lake water was 12°C in October and 23°C in May, what is the difference in degrees Fahrenheit?

 For more practice, visit booke.msscience.com/ math_practice

Open Oceans Life abounds in the open ocean. Scientists divide the ocean into different life zones, based on the depth to which sunlight penetrates the water. The lighted zone of the ocean is the upper 200 m or so. It is the home of the plankton that make up the foundation of the food chain in the open ocean. Below about 200 m is the dark zone of the ocean. Animals living in this region feed on material that floats down from the lighted zone, or they feed on each other. A few organisms are able to produce their own food.

Coral Reefs One of the most diverse ecosystems in the world is the coral reef. **Coral reefs** are formed over long periods of time from the calcium carbonate skeletons secreted by animals called corals. When corals die, their skeletons remain. Over time, the skeletal deposits form reefs such as the Great Barrier Reef off the coast of Australia, shown in **Figure 18.**

Reefs do not adapt well to long-term stress. Runoff from fields, sewage, and increased sedimentation from cleared land harm reef ecosystems. Organizations like the Environmental Protection Agency have developed management plans to protect the diversity of coral reefs. These plans treat a coral reef as a system that includes all the areas that surround the reef. Keeping the areas around reefs healthy will result in a healthy environment for the coral reef ecosystem.

Science Online

Topic: Coral Reefs
Visit booke.msscience.com for Web links to information about coral reef ecosystems.

Activity Construct a diorama of a coral reef. Include as many different kinds of organisms as you can for a coral reef ecosystem.

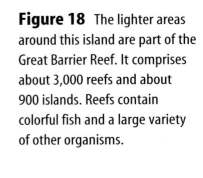

Figure 18 The lighter areas around this island are part of the Great Barrier Reef. It comprises about 3,000 reefs and about 900 islands. Reefs contain colorful fish and a large variety of other organisms.

Figure 19 As the tide recedes, small pools of seawater are left behind. These pools contain a variety of organisms such as sea stars and periwinkles.

Sea star

Periwinkles

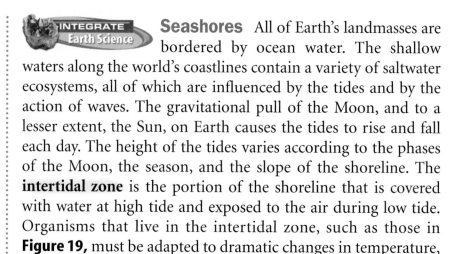

Seashores All of Earth's landmasses are bordered by ocean water. The shallow waters along the world's coastlines contain a variety of saltwater ecosystems, all of which are influenced by the tides and by the action of waves. The gravitational pull of the Moon, and to a lesser extent, the Sun, on Earth causes the tides to rise and fall each day. The height of the tides varies according to the phases of the Moon, the season, and the slope of the shoreline. The **intertidal zone** is the portion of the shoreline that is covered with water at high tide and exposed to the air during low tide. Organisms that live in the intertidal zone, such as those in **Figure 19,** must be adapted to dramatic changes in temperature, moisture, and salinity and must be able to withstand the force of wave action.

Estuaries Almost every river on Earth eventually flows into an ocean. The area where a river meets an ocean and contains a mixture of freshwater and salt water is called an **estuary** (ES chuh wer ee). Other names for estuaries include bays, lagoons, harbors, inlets, and sounds. They are located near coastlines and border the land. Salinity in estuaries changes with the amount of freshwater brought in by rivers and streams, and with the amount of salt water pushed inland by the ocean tides.

Estuaries, shown in **Figure 20,** are extremely fertile, productive environments because freshwater streams bring in tons of nutrients washed from inland soils. Therefore, nutrient levels in estuaries are higher than in freshwater ecosystems or other saltwater ecosystems.

Figure 20 The Chesapeake Bay is an estuary rich in resources. Fish and shrimp are harvested by commercial fishing boats.
Describe *what other resources can be found in estuaries.*

Estuary Life Organisms found in estuaries include many species of algae, salt-tolerant grasses, shrimp, crabs, clams, oysters, snails, worms, and fish. Estuaries also serve as important nurseries for many species of ocean fish. Estuaries provide much of the seafood consumed by humans.

✓ **Reading Check** *Why are estuaries more fertile than other aquatic ecosystems?*

section 3 review

Summary

Freshwater Ecosystems

- Temperature, light, salt, and dissolved oxygen are important factors.
- Rivers, streams, lakes, ponds, and wetlands are freshwater ecosystems.
- Human activities, such as too much lawn fertilizer, can pollute aquatic ecosystems.

Saltwater Ecosystems

- About 95 percent of Earth's water contains dissolved salts.
- Saltwater ecosystems include open oceans, coral reefs, seashores, and estuaries.
- Organisms that live on seashores have adaptations that enable them to survive dramatic changes in temperature, moisture, and salinity.
- Estuaries serve as nursery areas for many species of ocean fish.

Self Check

1. **Identify** the similarities and differences between a lake and a stream.
2. **Compare and contrast** the dark zone of the ocean with the forest floor of a tropical rain forest. What living or nonliving factors affect these areas?
3. **Explain** why fewer plants are at the bottom of deep lakes.
4. **Infer** what adaptations are necessary for organisms that live in the intertidal zone.
5. **Think Critically** Would you expect a fast moving mountain stream or the Mississippi River to have more dissolved oxygen? Explain.

Applying Skills

6. **Communicate** Wetlands trap and slowly release rain, snow, and groundwater. Describe in your Science Journal what might happen to a town located on a floodplain if nearby wetlands are destroyed.

Use the Internet

Explori\[\]g Wetlands

Goals

■ **Identify** wetland regions in the United States.

■ **Describe** the significance of the wetland ecosystem.

■ **Identify** plant and animal species native to a wetland region.

■ **Identify** strategies for supporting the preservation of wetlands.

Data Source

ScienceOnline

Visit **booke.msscience.com/** for more information about wetland environments and for data collected by other students.

⦿ *Real-World Question*

Wetlands, such as the one shown below, are an important part of the environment. These fertile ecosystems support unique plants and animals that can survive only in wetland conditions. The more you understand the importance of wetlands, the more you can do to preserve and protect them. Why are wetlands an important part of the ecosystem?

◉ Make a Plan

1. **Determine** where some major wetlands are located in the United States.
2. **Identify** one wetland area to study in depth. Where is it located? Is it classified as a marsh, bog, or something else?
3. **Explain** the role this ecosystem plays in the overall ecology of the area.
4. **Research information** about the plants and animals that live in the wetland environment you are researching.
5. **Investigate** what laws protect the wetland you are studying.

◉ Follow Your Plan

1. Make sure your teacher approves your plan before you start.
2. Perform the investigation.
3. Post your data at the link shown below.

◉ Analyze Your Data

1. **Describe** the wetland area you have researched. What region of the United States is it located in? What other ecological factors are found in that region?
2. **Outline** the laws protecting the wetland you are investigating. How long have the laws been in place?
3. **List** the plants and animals native to the wetland area you are researching. Are those plants and animals found in other parts of the region or the United States? What adaptations do the plants and animals have that help them survive in a wetland environment?

◉ Conclude and Apply

1. **Infer** Are all wetlands the same?
2. **Determine** what the ecological significance of the wetland area that you studied for that region of the country is.
3. **Draw Conclusions** Why should wetland environments be protected?
4. **Summarize** what people can do to support the continued preservation of wetland environments in the United States.

⬤ommunicating Your Data

Find this lab using the link below. **Post** your data in the table provided. **Review** other students' data to learn about other wetland environments in the United States.

Science⬤nline

booke.msscience.com/internet_lab

Creating Wetlands to Purify Wastewater

1 Pebbles were added to the Corrales wetlands to help with drainage.

2 Water irises thrived in the wetlands, less than a year after planting.

3 Students enjoy pure water from the Corrales wetlands after it is filtered.

When you wash your hands or flush the toilet, do you think about where the wastewater goes? In most places, it eventually ends up being processed in a traditional sewage-treatment facility. But some places are experimenting with a new method that processes wastewater by creating wetlands. Wetlands are home to filtering plants, such as cattails, and sewage-eating bacteria.

In 1996, school officials at the Corrales Elementary School in Albuquerque, New Mexico, faced a big problem. The old wastewater-treatment system had failed. Replacing it was going to cost a lot of money. Instead of constructing a new sewage-treatment plant, school officials decided to create a natural wetlands system. The wetlands system could do the job less expensively, while protecting the environment.

Today, this wetlands efficiently converts polluted water into cleaner water that's good for the environment. U.S. government officials are monitoring this alternative sewage-treatment system to see if it is successful. So far, so good!

Wetlands filter water through the actions of the plants and microorganisms that live there. When plants absorb water into their roots, some also take up pollutants. The plants convert the pollutants to forms that are not dangerous. At the same time, bacteria and other microorganisms are filtering water as they feed. Water moves slowly through wetlands, so the organisms have plenty of time to do their work. Wetlands built by people to filter small amounts of pollutants are called "constructed wetlands". In many places, constructed wetlands are better at cleaning wastewater than sewers or septic systems.

Visit and Observe Visit a wetlands and create a field journal of your observations. Draw the plants and animals you see. Use a field guide to help identify the wildlife. If you don't live near a wetlands, use resources to research wetlands environments.

Science Online

For more information, visit booke.msscience.com/time

Reviewing Main Ideas

Section 1 **How Ecosystems Change**

1. Ecological succession is the gradual change from one plant community to another.

2. Primary succession begins in a place where no plants were before.

3. Secondary succession begins in a place that has soil and was once the home of living organisms.

4. A climax community has reached a stable stage of ecological succession.

Section 2 **Biomes**

1. Temperature and precipitation help determine the climate of a region.

2. Large geographic areas with similar climax communities are called biomes.

3. Earth's land biomes include tundra, taiga, temperate deciduous forest, temperate rain forest, tropical rain forest, grassland, and desert.

Section 3 **Aquatic Ecosystems**

1. Freshwater ecosystems include streams, rivers, lakes, ponds, and wetlands.

2. Wetlands are areas that are covered with water most of the year. They are found in regions that lie between land-masses and water.

3. Saltwater ecosystems include estuaries, sea-shores, coral reefs, a few inland lakes, and the deep ocean.

4. Estuaries are fertile transitional zones between freshwater and saltwater environments.

Visualizing Main Ideas

Copy and complete this concept map about land biomes.

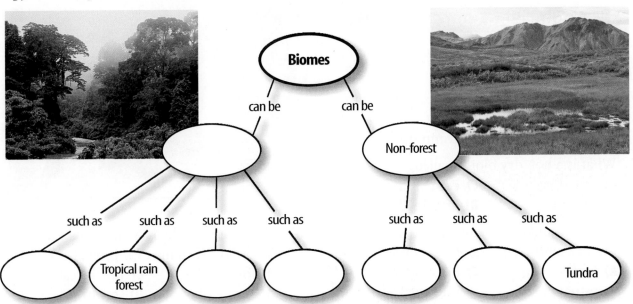

Using Vocabulary

biome p. 68
climax community p. 67
coral reef p. 81
desert p. 74
estuary p. 82
grassland p. 75
intertidal zone p. 82
pioneer species p. 64

succession p. 64
taiga p. 70
temperate deciduous
    forest p. 71
temperate rain forest p. 71
tropical rain forest p. 72
tundra p. 69
wetland p. 79

Fill in the blanks with the correct vocabulary word or words.

1. _____ refers to the normal changes in the types of species that live in communities.

2. A(n) _____ is a group of organisms found in a stable stage of succession.

3. Deciduous trees are dominant in the _____.

4. The average temperature in _____ is between 9°C and 12°C.

5. _____ are the most biologically diverse biomes in the world.

6. A(n) _____ is an area where freshwater meets the ocean.

Checking Concepts

Choose the word or phrase that best answers the question.

7. What are tundra and desert examples of?
   A) ecosystems
   B) biomes
   C) habitats
   D) communities

8. What is a hot, dry biome called?
   A) desert
   B) tundra
   C) coral reef
   D) grassland

9. Where would organisms that are adapted to live in slightly salty water be found?
   A) lake
   B) estuary
   C) open ocean
   D) intertidal zone

10. Which biome contains mostly frozen soil called permafrost?
    A) taiga
    B) temperate rain forest
    C) tundra
    D) temperate deciduous forest

11. A new island is formed from a volcanic eruption. Which species probably would be the first to grow and survive?
    A) palm trees
    B) lichens
    C) grasses
    D) ferns

12. What would the changes in communities that take place on a recently formed volcanic island best be described as?
    A) primary succession
    B) secondary succession
    C) tertiary succession
    D) magma

13. What is the stable end stage of succession?
    A) pioneer species
    B) climax community
    C) limiting factor
    D) permafrost

Use the illustration below to answer question 14.

Observed Fire Danger Class—June, 2003

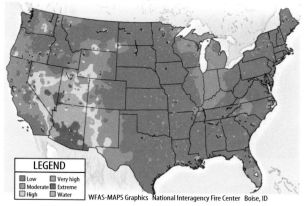

LEGEND
■ Low   ■ Very high
■ Moderate ■ Extreme
■ High   ■ Water

WFAS-MAPS Graphics National Interagency Fire Center Boise, ID

14. Which area of the U.S. had the highest observed fire danger on June 20, 2003?
    A) northeast
    B) southeast
    C) northwest
    D) southwest

Science Online booke.msscience.com/vocabulary_puzzlemaker

Thinking Critically

15. **Explain** In most cases, would a soil sample from a temperate deciduous forest be more or less nutrient-rich than a soil sample from a tropical rain forest?

16. **Explain** why some plant seeds need fire in order to germinate. How does this give these plants an advantage in secondary succession?

17. **Determine** A grassy meadow borders a beech-maple forest. Is one of these ecosystems undergoing succession? Why?

18. **Infer** why tundra plants are usually small.

19. **Make and Use a Table** Copy and complete the following table about aquatic ecosystems. Include these terms: *intertidal zone, lake, pond, coral reef, open ocean, river, estuary,* and *stream.*

Aquatic Ecosystems	
Saltwater	**Freshwater**
Do not write in this book.	

20. **Recognize Cause and Effect** Wildfires like the one in Yellowstone National Park in 1988, cause many changes to the land. Determine the effect of a fire on an area that has reached its climax community.

Performance Activities

21. **Oral Presentation** Research a biome not in this chapter. Find out about its climate and location, and which organisms live there. Present this information to your class.

Applying Math

Use the table below to answer question 22.

Rainfall Amounts	
Biome	**Average Precipitation/Year (cm)**
Taiga	50
Temperate rain forest	200
Tropical rain forest	400
Desert	25
Temperate deciduous forest	150
Tundra	25

22. **Biome Precipitation** How many times more precipitation does the tropical rain forest biome receive than the taiga or desert?

Use the graph below to answer question 23.

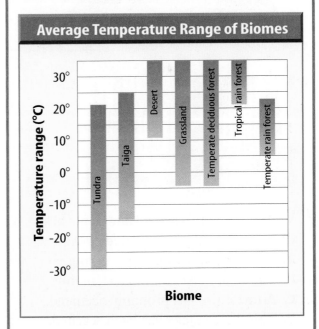

Average Temperature Range of Biomes

23. **Biome Temperatures** According to the graph, which biome has the greatest and which biome has the least variation in temperature throughout the year? Estimate the difference between the two.

Part 1 | Multiple Choice

Record your answers on the answer sheet provided by your teacher or on a sheet of paper.

1. What two factors are most responsible for limiting life in a particular area?
   A. sunlight and temperature
   B. precipitation and temperature
   C. precipitation and sunlight
   D. soil conditions and precipitation

2. Which of the following forms during primary succession?
   A. trees   C. wildlife
   B. soil   D. grasses

Use the illustrations below to answer questions 3 and 4.

A   Lichens

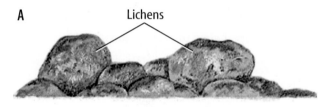

B

3. Which of the following statements best describes what is represented by A?
   A. Primary succession is occurring.
   B. Secondary succession is occurring.
   C. A forest fire has probably occurred.
   D. The climax stage has been reached.

4. Which of the following statements best describes what is represented by B?
   A. The climax stage has been reached.
   B. Pioneer species are forming soil.
   C. Bare rock covers most of the area.
   D. Secondary succession is occurring.

Use the map below to answer questions 5 and 6.

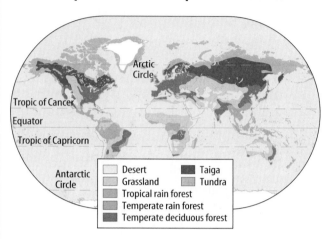

Desert	Taiga
Grassland	Tundra
Tropical rain forest	
Temperate rain forest	
Temperate deciduous forest	

5. What biome is located in the latitudes just south of the north pole?
   A. taiga
   B. temperate deciduous rain forest
   C. tundra
   D. temperate rain forest

6. The tropical rainforest biome is found primarily near the
   A. Arctic Circle.
   B. Tropic of Cancer.
   C. equator.
   D. Tropic of Capricorn.

7. Which of the following is composed of a mix of salt water and freshwater?
   A. an intertidal zone
   B. an estuary
   C. a seashore
   D. a coral reef

Test-Taking Tip

Come Prepared Bring at least two sharpened No. 2 pencils and a good eraser to the test. Check to make sure that your eraser completely removes all pencil marks.

Part 2 | Short Response/Grid In

Record your answers on the answer sheet provided by your teacher or on a sheet of paper.

8. Name two products that come from wetlands.

9. Which takes longer, primary succession or secondary succession? Why?

Use the photos below to answer questions 10 and 11.

10. In what biome would you most likely find A? How is this animal adapted to survive in its biome?

11. In what biome would you most likely find B? How is this animal adapted to survive in its biome?

12. Which biome receives the most rainfall per year? Which receives the least rainfall?

13. Why are forests unlikely to develop in grasslands?

14. What are two kinds of wetlands? What kinds of animals are found in wetlands?

15. What organisms inhabit the upper zone of the open ocean and why are they so important?

Part 3 | Open Ended

Record your answers on a sheet of paper.

16. Explain how lichens contribute to the process of soil formation.

17. Compare and contrast a freshwater lake ecosystem with a freshwater pond ecosystem.

18. What special adaptations must all of the organisms that live in the intertidal zone have?

19. What are the differences between the temperate rain forest biome and the tropical rain forest biome?

Use the illustration below to answer questions 20 and 21.

20. Identify and describe zone C in the diagram. What kinds of wildlife are found there?

21. Identify zone D and zone A. Describe the environment in each zone. Why might an organism that lives in zone A not be able to survive in zone D?

22. Discuss the effects of human impact on freshwater environments like lakes and ponds.

The BIG Idea

Many of Earth's resources are limited.

SECTION 1
Resources
Main Idea Earth has natural resources that can be replenished as well as natural resources that cannot be replenished.

SECTION 2
Pollution
Main Idea Air, water, and soil pollution have many causes, including hazardous waste and the burning of fossil fuels.

SECTION 3
The Three Rs of Conservation
Main Idea Natural resources can be conserved by following the three Rs of conservation: reduce, reuse, and recycle.

Conserving Resources

Resources Fuel Our Lives

Resources, such as clean water and air, are commonly taken for granted. We depend on water and air to survive. Fossil fuels are another type of resource, and we depend on them for energy. However, fossil fuels can pollute our air and water.

Science Journal List some other resources that we depend on and describe how we use them.

Start-Up Activities

What happens when topsoil is left unprotected?

Plants grow in the top, nutrient-rich layer, called topsoil. Plants help keep topsoil in place by protecting it from wind and rain. Try the following experiment to find out what happens when topsoil is left unprotected.

1. Use a mixture of moist sand and potting soil to create a miniature landscape in a plastic basin or aluminum-foil baking pan. Form hills and valleys in your landscape.

2. Use clumps of moss to cover areas of your landscape. Leave some sloping portions without plant cover.

3. Simulate a rainstorm over your landscape by spraying water on it from a spray bottle or by pouring a slow stream of water on it from a beaker.

4. **Think Critically** In your Science Journal, record your observations and describe what happened to the land that was not protected by plant cover.

 **Resources** Make the following Foldable to help you organize information and diagram ideas about renewable and nonrenewable resources.

STEP 1 **Fold** a sheet of paper in half lengthwise. Make the back edge about 5 cm longer than the front edge.

STEP 2 **Turn** the paper so the fold is on the bottom. Then **fold** in half.

STEP 3 **Unfold and cut** only the top layer along the fold to make two tabs. **Label** the Foldable as shown.

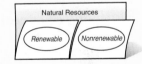

Make a Concept Map Before you read the chapter, list examples of each type of natural resource you already know on the back of the appropriate tabs. As you read the chapter, add to your lists.

Science Online Preview this chapter's content and activities at booke.mrscience.com

Get Ready to Read

① Learn It! Knowing how to find answers to questions will help you on reviews and tests. Some answers can be found in the textbook, while other answers require you to go beyond the textbook. These answers might be based on knowledge you already have or things you have experienced.

② Practice It! Read the excerpt below. Answer the following questions and then discuss them with a partner.

> Even though renewable resources are recycled or replaced, they are sometimes in short supply. Rain and melted snow replace the water in streams, lakes, and reservoirs. Sometimes there may not be enough rain or snowmelt to meet all the needs of people, plants, and animals. In some parts of the world, especially desert regions, water and other resources usually are scarce. Other resources can be used instead.
>
> —*from page 95*

- How is water in streams, lakes, and reservoirs replenished?
- What happens if there is not enough rain or snowmelt to replace water that has been used?
- How can you conserve natural resources, such as water?

③ Apply It! Look at some questions in the text. Which questions can be answered directly from the text? Which require you to go beyond the text?

Target Your Reading

Reading Tip

As you read, keep track of questions you answer in the chapter. This will help you remember what you read.

Use this to focus on the main ideas as you read the chapter.

① **Before you read** the chapter, respond to the statements below on your worksheet or on a numbered sheet of paper.

- Write an **A** if you **agree** with the statement.
- Write a **D** if you **disagree** with the statement.

② **After you read** the chapter, look back to this page to see if you've changed your mind about any of the statements.

- If any of your answers changed, explain why.
- Change any false statements into true statements.
- Use your revised statements as a study guide.

Science Online

Print out a worksheet of this page at booke.msscience.com

Before You Read A or D		Statement	After You Read A or D
	1	All Earth's resources are able to be replenished.	
	2	Fossil fuels must be burned to release the energy they hold.	
	3	The only unlimited source of energy for Earth is the Sun.	
	4	Air pollution can be cleaned when it reacts with sunlight.	
	5	Underground water supplies are safe from pollution.	
	6	Some household items, such as batteries and paint, are hazardous and may pollute the environment if disposed incorrectly.	
	7	Conservation can help prevent shortages of natural resources.	
	8	Certain plastics can be recycled into items such as carpeting and clothing.	
	9	Aluminum is the only metal that can be recycled.	

Resources

What You'll Learn

- **Compare** renewable and nonrenewable resources.
- **List** uses of fossil fuels.
- **Identify** alternatives to fossil fuel use.

Why It's Important

Wise use of natural resources is important for the health of all life on Earth.

🔍 Review Vocabulary

geyser: a spring that emits intermittent jets of heated water and steam

New Vocabulary

- natural resource
- renewable resource
- nonrenewable resource
- petroleum
- fossil fuel
- hydroelectric power
- nuclear energy
- geothermal energy

Natural Resources

An earthworm burrowing in moist soil eats decaying plant material. A robin catches the worm and flies to a tree. The leaves of the tree use sunlight during photosynthesis. Leaves fall to the ground, decay, and perhaps become an earthworm's meal. What do these living things have in common? They rely on Earth's **natural resources**—the parts of the environment that are useful or necessary for the survival of living organisms.

What kinds of natural resources do you use? Like other organisms, you need food, air, and water. You also use resources that are needed to make everything from clothes to cars. Natural resources supply energy for automobiles and power plants. Although some natural resources are plentiful, others are not.

Renewable Resources   The Sun, an inexhaustible resource, provides a constant supply of heat and light. Rain fills lakes and streams with water. When plants carry out photosynthesis, they add oxygen to the air. Sunlight, water, air, and the crops shown in **Figure 1** are examples of renewable resources. A **renewable resource** is any natural resource that is recycled or replaced constantly by nature.

Figure 1 Cotton and wood are renewable resources. Cotton cloth is used for rugs, curtains, and clothing. A new crop of cotton can be grown every year. Wood is used for furniture, building materials, and paper. It will take 20 years for these young trees to grow large enough to harvest.

Cotton plants

Tree farm

Supply and Demand Even though renewable resources are recycled or replaced, they are sometimes in short supply. Rain and melted snow replace the water in streams, lakes, and reservoirs. Sometimes, there may not be enough rain or snowmelt to meet all the needs of people, plants, and animals. In some parts of the world, especially desert regions, water and other resources usually are scarce. Other resources can be used instead, as shown in **Figure 2.**

Nonrenewable Resources Natural resources that are used up more quickly than they can be replaced by natural processes are **nonrenewable resources.** Earth's supply of nonrenewable resources is limited. You use nonrenewable resources when you take home groceries in a plastic bag, paint a wall, or travel by car. Plastics, paints, and gasoline are made from an important nonrenewable resource called petroleum, or oil. **Petroleum** is formed mostly from the remains of microscopic marine organisms buried in Earth's crust. It is nonrenewable because it takes hundreds of millions of years for it to form.

✔️ **Reading Check** *What are nonrenewable resources?*

Minerals and metals found in Earth's crust are nonrenewable resources. Petroleum is a mineral. So are diamonds and the graphite in pencil lead. The aluminum used to make soft-drink cans is a metal. Iron, copper, tin, gold, silver, tungsten, and uranium also are metals. Many manufactured items, like the car shown in **Figure 3,** are made from nonrenewable resources.

Figure 2 In parts of Africa, fire-wood has become scarce. People in this village now use solar energy instead of wood for cooking.

Figure 3 Iron, a nonrenewable resource, is the main ingredient in steel. Steel is used to make cars, trucks, appliances, buildings, bridges, and even tires.
Infer *what other nonrenewable resources are used to build a car.*

Mini LAB

Observing Mineral Mining Effects

Procedure 🚫 🧤 🗑️

1. Place a **chocolate-chip cookie** on a **paper plate.** Pretend the chips are mineral deposits and the rest of the cookie is Earth's crust.
2. Use a **toothpick** to locate and dig up the mineral deposits. Try to disturb the land as little as possible.
3. When mining is completed, try to restore the land to its original condition.

Analysis

1. How well were you able to restore the land?
2. Compare the difficulty of digging for mineral deposits found close to the surface with digging for those found deep in Earth's crust.
3. Describe environmental changes that might result from a mining operation.

Try at Home

Fossil Fuels

Coal, oil, and natural gas are nonrenewable resources that supply energy. Most of the energy you use comes from these fossil fuels, as the graph in **Figure 4** shows. **Fossil fuels** are fuels formed in Earth's crust over hundreds of millions of years. Cars, buses, trains, and airplanes are powered by gasoline, diesel fuel, and jet fuel, which are made from oil. Coal is used in many power plants to produce electricity. Natural gas is used in manufacturing, for heating and cooking, and sometimes as a vehicle fuel.

Fossil Fuel Conservation Billions of people all over the world use fossil fuels every day. Because fossil fuels are nonrenewable, Earth's supply of them is limited. In the future, they may become more expensive and difficult to obtain. Also, the use of fossil fuels can lead to environmental problems. For example, mining coal can require stripping away thick layers of soil and rock, as shown in **Figure 4,** which destroys ecosystems. Another problem is that fossil fuels must be burned to release the energy stored in them. The burning of fossil fuels produces waste gases that cause air pollution, including smog and acid rain. For these reasons, many people suggest reducing the use of fossil fuels and finding other sources of energy.

You can use simple conservation measures to help reduce fossil fuel use. Switch off the light when you leave a room and turn off the television when you're not watching it. These actions reduce your use of electricity, which often is produced in power plants that burn fossil fuels. Hundreds of millions of automobiles are in use in the United States. Riding in a car pool or taking public transportation uses fewer liters of gasoline than driving alone in a car. Walking or riding a bicycle uses even less fossil fuel. Reducing fossil fuel use has an added benefit—the less you use, the more money you save.

Figure 4 Coal is a fossil fuel. It often is obtained by strip mining, which removes all the soil above the coal deposit. The soil is replaced, but it takes many years for the ecosystem to recover.

Identify *the resource that provided 84 percent of the energy used in the United States in 1999.*

Renewable energy

Nuclear power 8%

Renewable energy 8%

Oil 39%

Natural gas 23%

Coal 22%

Figure 5 Most power plants use turbine generators to produce electricity. In fossil fuel plants, burning fuel boils water and produces steam that turns the turbine.

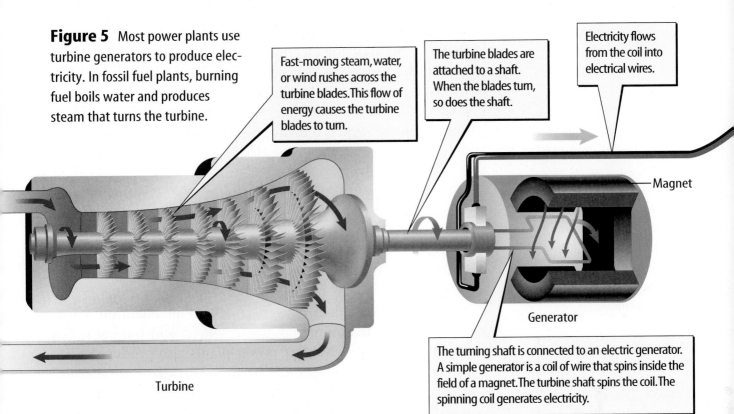

Fast-moving steam, water, or wind rushes across the turbine blades. This flow of energy causes the turbine blades to turn.

The turbine blades are attached to a shaft. When the blades turn, so does the shaft.

Electricity flows from the coil into electrical wires.

Magnet

Generator

The turning shaft is connected to an electric generator. A simple generator is a coil of wire that spins inside the field of a magnet. The turbine shaft spins the coil. The spinning coil generates electricity.

Turbine

Alternatives to Fossil Fuels

Another approach to reducing fossil fuel use is to develop other sources of energy. Much of the electricity used today comes from power plants that burn fossil fuels. As **Figure 5** shows, electricity is generated when a rotating turbine turns a coil of wires in the magnetic field of an electric generator. Fossil-fuel power plants boil water to produce steam that turns the turbine. Alternative energy sources, including water, wind, and atomic energy can be used instead of fossil fuels to turn turbines. Also, solar cells can produce electricity using only sunlight, with no turbines at all. Some of these alternative energy sources—particularly wind and solar energy—are so plentiful they could be considered inexhaustible resources.

Water Power Water is a renewable energy source that can be used to generate electricity. **Hydroelectric power** is electricity that is produced when the energy of falling water is used to turn the turbines of an electric generator. Hydroelectric power does not contribute to air pollution because no fuel is burned. However, it does present environmental concerns. Building a hydroelectric plant usually involves constructing a dam across a river. The dam raises the water level high enough to produce the energy required for electricity generation. Many acres behind the dam are flooded, destroying land habitats and changing part of the river into a lake.

INTEGRATE Social Studies

Energy Oil and natural gas are used to produce over 60 percent of the energy supply in the United States. Over half of the oil used is imported from other countries. Many scientists suggest that emissions from the burning of fossil fuels are principally responsible for global warming. In your Science Journal, write what you might do to persuade utility companies to increase their use of water, wind, and solar power.

Wind Power Wind power is another renewable energy source that can be used for electricity production. Wind turns the blades of a turbine, which powers an electric generator. When winds blow at least 32 km/h, energy is produced. Wind power does not cause air pollution, but electricity can be produced only when the wind is blowing. So far, wind power accounts for only a small percentage of the electricity used worldwide.

Nuclear Power Another alternative to fossil fuels makes use of the huge amounts of energy in the nuclei of atoms, as shown in **Figure 6. Nuclear energy** is released when billions of atomic nuclei from uranium, a radioactive element, are split apart in a nuclear fission reaction. This energy is used to produce steam that rotates the turbine blades of an electric generator.

Nuclear power does not contribute to air pollution. However, uranium is a nonrenewable resource, and mining it can disrupt ecosystems. Nuclear power plants also produce radioactive wastes that can seriously harm living organisms. Some of these wastes remain radioactive for thousands of years, and their safe disposal is a problem that has not yet been solved. Accidents also are a danger.

Figure 6 Nuclear power plants are designed to withstand the high energy produced by nuclear reactions.
Describe *how heat is produced in a nuclear reactor.*

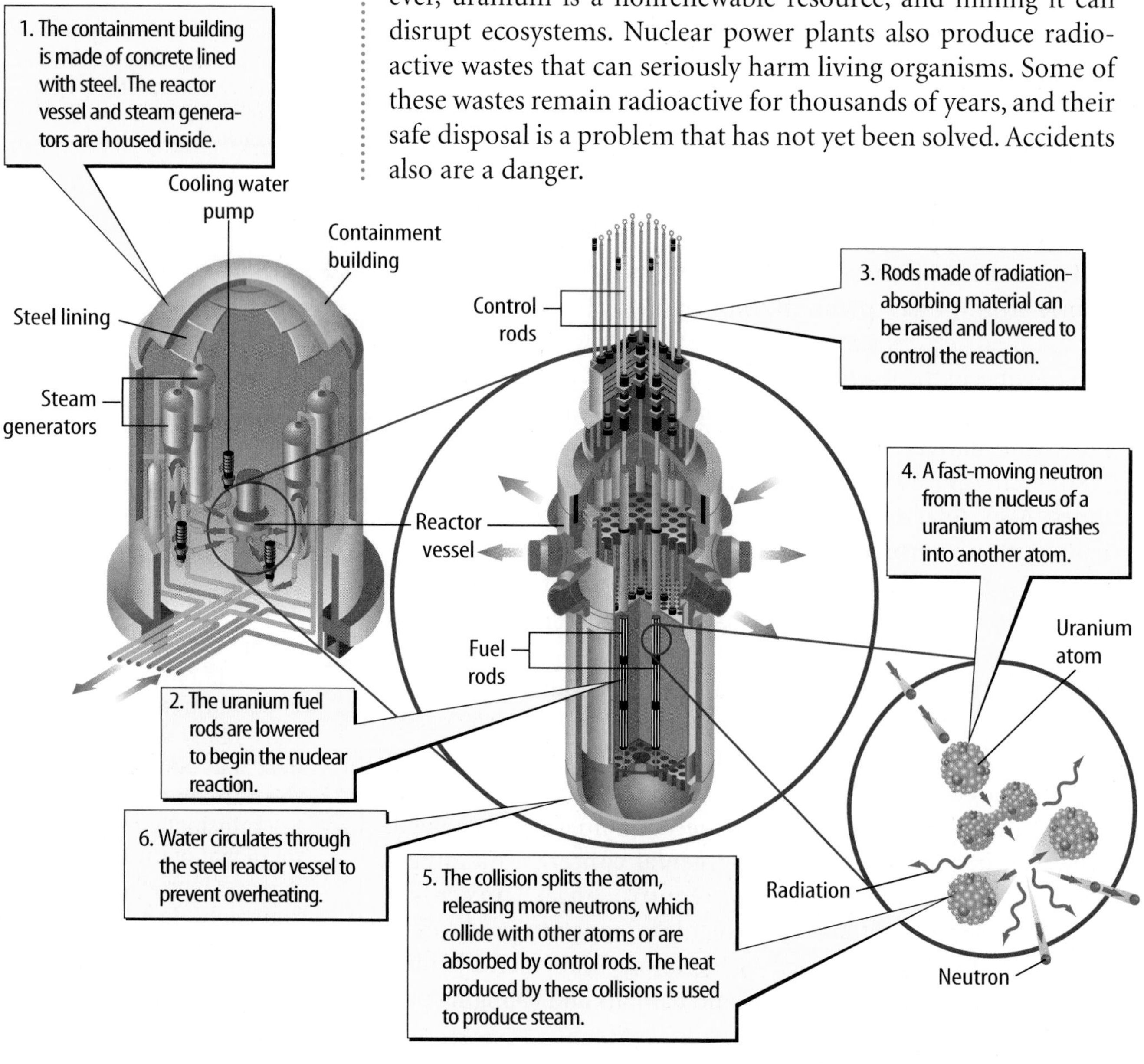

1. The containment building is made of concrete lined with steel. The reactor vessel and steam generators are housed inside.

Cooling water pump

Containment building

Steel lining

Control rods

3. Rods made of radiation-absorbing material can be raised and lowered to control the reaction.

Steam generators

4. A fast-moving neutron from the nucleus of a uranium atom crashes into another atom.

Reactor vessel

Uranium atom

Fuel rods

2. The uranium fuel rods are lowered to begin the nuclear reaction.

6. Water circulates through the steel reactor vessel to prevent overheating.

5. The collision splits the atom, releasing more neutrons, which collide with other atoms or are absorbed by control rods. The heat produced by these collisions is used to produce steam.

Radiation

Neutron

Geothermal Energy

The hot, molten rock that lies deep beneath Earth's surface is also a source of energy. You see the effects of this energy when lava and hot gases escape from an erupting volcano or when hot water spews from a geyser. The heat energy contained in Earth's crust is called **geothermal energy.** Most geothermal power plants use this energy to produce steam to generate electricity.

Geothermal energy for power plants is available only where natural geysers or volcanoes are found. A geothermal power plant in California uses steam produced by geysers. The island nation of Iceland was formed by volcanoes, and geothermal energy is plentiful there. Geothermal power plants supply heat and electricity to about 90 percent of the homes in Iceland. Outdoor swimming areas also are heated with geothermal energy, as shown in **Figure 7.**

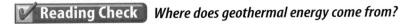 **Reading Check** *Where does geothermal energy come from?*

Solar Energy

The most inexhaustible source of energy for all life on Earth is the Sun. Solar energy is an alternative to fossil fuels. One use of solar energy is in solar-heated buildings. During winter in the northern hemisphere, the parts of a building that face south receive the most sunlight. Large windows placed on the south side of a building help heat it by allowing warm sunshine into the building during the day. Floors and walls of most solar-heated buildings are made of materials that absorb heat during the day. During the night, the stored heat is released slowly, keeping the building warm. **Figure 8** shows how solar energy can be used.

Figure 7 In Iceland, a geothermal power plant pumps hot water out of the ground to heat buildings and generate electricity. Leftover hot water goes into this lake, making it warm enough for swimming even when the ground is covered with snow.

Figure 8 The Zion National Park Visitor Center in Utah is a solar-heated building designed to save energy. The roof holds solar panels that are used to generate electricity. High windows can be opened to circulate air and help cool the building on hot days. The overhanging roof shades the windows during summer.

Free electrons

Layers of semiconductor atoms

Electric current flows through the calculator and back to the PV cell to form a complete circuit.

Light

Figure 9 Light energy from the Sun travels in tiny packets of energy called photons. Photons crash into the atoms of PV cells, knocking electrons loose. These electrons create an electric current.

Solar Cells Do you know how a solar-powered calculator works? How do spacecraft use sunlight to generate electricity? These devices use photovoltaic (foh toh vohl TAY ihk) cells to turn sunlight into electric current, as shown in **Figure 9.** Photovoltaic (PV) cells are small and easy to use. However, they produce electricity only in sunlight, so batteries are needed to store electricity for use at night or on cloudy days. Also, PV cells presently are too expensive to use for generating large amounts of electricity. Improvements in this technology continue to be made, and prices probably will go down in the future. As **Figure 10** shows, solar buildings and PV cells are just two of the many ways solar energy can be used to replace fossil fuels.

section ① review

Summary

Natural Resources

- All living things depend on natural resources to survive.
- Some resources are renewable, while other resources, such as petroleum, are nonrenewable.

Fossil Fuels

- Most of the energy that humans use comes from fossil fuels.
- Fossil fuels must be burned to release the energy stored in them, which causes air pollution.

Alternatives to Fossil Fuels

- Alternatives to fossil fuels include water power, wind power, nuclear power, geothermal energy, and solar energy.
- The Sun provides the most inexhaustible supply of energy for all life on Earth.

Self Check

1. **Summarize** What are natural resources?
2. **Compare and contrast** renewable and nonrenewable resources. Give five examples of each.
3. **Describe** the advantages and disadvantages of using nuclear power.
4. **Describe** two ways solar energy can be used to reduce fossil fuel use.
5. **Think Critically** Explain why the water that is used to cool the reactor vessel of a nuclear power plant is kept separate from the water that is heated to produce steam for the turbine generators.

Applying Math

6. **Solve One-Step Equations** Most cars in the U.S. are driven about 10,000 miles each year. If a car can travel 30 miles on one gallon of gasoline, how many gallons will it use in a year?

Figure 10

Sunlight is a renewable energy source that provides an alternative to fossil fuels. Solar technologies use the Sun's energy in many ways—from heating to electricity generation.

▼ **ELECTRICITY** Photovoltaic (PV) cells turn sunlight into electric current. They are commonly used to power small devices, such as calculators. Panels that combine many PV cells provide enough electricity for a home—or an orbiting satellite, such as the International Space Station, below.

▲ **POWER PLANTS** In California's Mojave Desert, an experimental solar power plant used hundreds of mirrors to focus sunlight on a water-filled tower. The steam produced by this system generates enough electricity to power 2,400 homes.

▼ **COOKING** In hot, sunny weather, a solar oven or panel cooker can be used to cook a pot of rice or heat water. The powerful solar cooker shown below reaches even higher temperatures. It is being used to fry food.

▼ **INDOOR HEATING** South-facing windows and heat-absorbing construction materials turn a room into a solar collector that can help heat an entire building, such as this Connecticut home.

▲ **WATER HEATING** Water is heated as it flows through small pipes in this roof-mounted solar heat collector. The hot water then flows into an insulated tank for storage.

Pollution

as you read

What You'll Learn

- **Describe** types of air pollution.
- **Identify** causes of water pollution.
- **Explain** methods that can be used to prevent erosion.

Why It's Important

By understanding the causes of pollution, you can help solve pollution problems.

🔍 Review Vocabulary

atmosphere: the whole mass of air surrounding Earth

New Vocabulary

- pollutant
- acid precipitation
- greenhouse effect
- ozone depletion
- erosion
- hazardous waste

Keeping the Environment Healthy

More than six billion people live on Earth. This large human population puts a strain on the environment, but each person can make a difference. You can help safeguard the environment by paying attention to how your use of natural resources affects air, land, and water.

Air Pollution

On a still, sunny day in almost any large city, you might see a dark haze in the air, like that in **Figure 11.** The haze comes from pollutants that form when wood or fuels are burned. A **pollutant** is a substance that contaminates the environment. Air pollutants include soot, smoke, ash, and gases such as carbon dioxide, carbon monoxide, nitrogen oxides, and sulfur oxides. Wherever cars, trucks, airplanes, factories, homes, or power plants are found, air pollution is likely. Air pollution also can be caused by volcanic eruptions, wind-blown dust and sand, forest fires, and the evaporation of paints and other chemicals.

Smog is a form of air pollution created when sunlight reacts with pollutants produced by burning fuels. It can irritate the eyes and make breathing difficult for people with asthma or other lung diseases. Smog can be reduced if people take buses or trains instead of driving or if they use vehicles, such as electric cars, that produce fewer pollutants than gasoline-powered vehicles.

Figure 11 The term *smog* was used for the first time in the early 1900s to describe the mixture of smoke and fog that often covers large cities in the industrial world. **Infer** *how smog can be reduced in large cities.*

Figure 12 Compare these two photographs of the same statue. The photo on the left was taken before acid rain became a problem. The photo on the right shows acid rain damage. The pH scale, shown below, indicates whether a solution is acidic or basic.

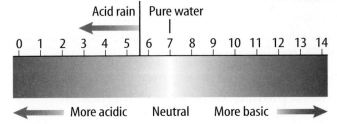

Acid Precipitation

Water vapor condenses on dust particles in the air to form droplets that combine to create clouds. Eventually, the droplets become large enough to fall to the ground as precipitation—mist, rain, snow, sleet, or hail. Air pollutants from the burning of fossil fuels can react with water in the atmosphere to form strong acids. Acidity is measured by a value called pH, as shown in **Figure 12. Acid precipitation** has a pH below 5.6.

Effects of Acid Rain Acid precipitation washes nutrients from the soil, which can lead to the death of trees and other plants. Runoff from acid rain that flows into a lake or pond can lower the pH of the water. If algae and microscopic organisms cannot survive in the acidic water, fish and other organisms that depend on them for food also die.

Preventing Acid Rain Sulfur from burning coal and nitrogen oxides from vehicle exhaust are the pollutants primarily responsible for acid rain. Using low-sulfur fuels, such as natural gas or low-sulfur coal, can help reduce acid precipitation. However, these fuels are less plentiful and more expensive than high-sulfur coal. Smokestacks that remove sulfur dioxide before it enters the atmosphere also help. Reducing automobile use and keeping car engines properly tuned can reduce acid rain caused by nitrogen oxide pollution. The use of electric cars, or hybrid-fuel cars that can run on electricity as well as gasoline, also could help.

Mini LAB

Measuring Acid Rain

Procedure
1. Collect **rainwater** by placing a clean **cup** outdoors. Do not collect rainwater that has been in contact with any object or organism.
2. Dip a piece of **pH indicator paper** into the sample.
3. Compare the color of the paper to the pH chart provided. Record the pH of the rainwater.
4. Use separate pieces of pH paper to test the pH of **tap water** and **distilled water**. Record these results.

Analysis
1. Is the rainwater acidic, basic, or neutral?
2. How does the pH of the rainwater compare with the pH of tap water? With the pH of distilled water?

Greenhouse Effect

Sunlight travels through the atmosphere to Earth's surface. Some of this sunlight normally is reflected back into space. The rest is trapped by certain atmospheric gases, as shown in **Figure 13.** This heat-trapping feature of the atmosphere is the **greenhouse effect.** Without it, temperatures on Earth probably would be too cold to support life.

Atmospheric gases that trap heat are called greenhouse gases. One of the most important greenhouse gases is carbon dioxide (CO_2). CO_2 is a normal part of the atmosphere. It is also a waste product that forms when fossil fuels are burned. Over the past century, more fossil fuels have been burned than ever before, which is increasing the percentage of CO_2 in the atmosphere. The atmosphere might be trapping more of the Sun's heat, making Earth warmer. A rise in Earth's average temperature, possibly caused by an increase in greenhouse gases, is known as global warming.

Global Warming Temperature data collected from 1895 through 1995 indicate that Earth's average temperature increased about 1°C during that 100-year period. No one is certain whether this rise was caused by human activities or is a natural part of Earth's weather cycle. What kinds of changes might be caused by global warming? Changing rainfall patterns could alter ecosystems and affect the kinds of crops that can be grown in different parts of the world. The number of storms and hurricanes might increase. The polar ice caps might begin to melt, raising sea levels and flooding coastal areas. Warmer weather might allow tropical diseases, such as malaria, to become more widespread. Many people feel that the possibility of global warming is a good reason to reduce fossil fuel use.

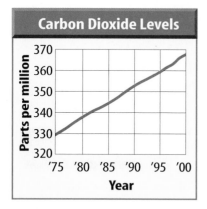

Figure 13 The moment you step inside a greenhouse, you feel the results of the greenhouse effect. Heat trapped by the glass walls warms the air inside. In a similar way, atmospheric greenhouse gases trap heat close to Earth's surface.

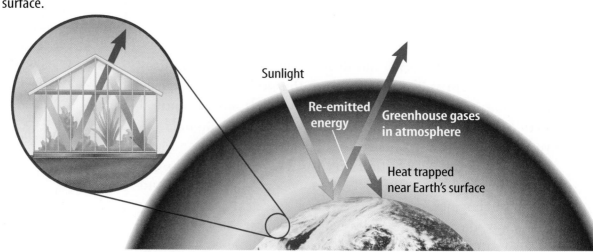

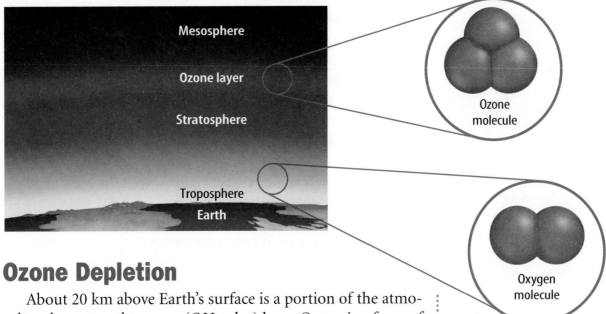

Ozone Depletion

About 20 km above Earth's surface is a portion of the atmosphere known as the ozone (OH zohn) layer. Ozone is a form of oxygen, as shown in **Figure 14.** The ozone layer absorbs some of the Sun's harmful ultraviolet (UV) radiation. UV radiation can damage living cells.

Every year, the ozone layer temporarily becomes thinner over each polar region during its spring season. The thinning of the ozone layer is called **ozone depletion.** This problem is caused by certain pollutant gases, especially chlorofluorocarbons (klor oh FLOR oh kar bunz) (CFCs). CFCs are used in the cooling systems of refrigerators, freezers, and air conditioners. When CFCs leak into the air, they slowly rise into the atmosphere until they arrive at the ozone layer. CFCs react chemically with ozone, breaking apart the ozone molecules.

UV Radiation Because of ozone depletion, the amount of UV radiation that reaches Earth's surface could be increasing. UV radiation could be causing a rise in the number of skin cancer cases in humans. It also might be harming other organisms. The ozone layer is so important to the survival of life on Earth that world governments and industries have agreed to stop making and using CFCs.

Ozone that is high in the upper atmosphere protects life on Earth. Near Earth's surface though, it can be harmful. Ozone is produced when fossil fuels are burned. This ozone stays in the lower atmosphere, where it pollutes the air. Ozone damages the lungs and other sensitive tissues of animals and plants. For example, it can cause the needles of a Ponderosa pine to drop, harming growth.

Figure 14 The atmosphere's ozone layer absorbs large amounts of UV radiation, preventing it from reaching Earth's surface. Ozone molecules are made of three oxygen atoms. They are formed in a chemical reaction between sunlight and oxygen. The oxygen you breathe has two oxygen atoms in each molecule.
Infer *what will happen if the ozone layer continues to thin.*

☑ Reading Check *What is the difference between ozone in the upper atmosphere and ozone in the lower atmosphere?*

Air Quality Carbon monoxide enters the body through the lungs. It attaches to red blood cells, preventing the cells from absorbing oxygen. In your Science Journal, explain why heaters and barbecues designed for outdoor use never should be used indoors.

Indoor Air Pollution

Air pollution can occur indoors. Today's buildings are better insulated to conserve energy. However, better insulation reduces the flow of air into and out of a building, so air pollutants can build up indoors. For example, burning cigarettes release hazardous particles and gases into the air. Even nonsmokers can suffer ill effects from secondhand cigarette smoke. As a result, smoking no longer is allowed in many public and private buildings. Paints, carpets, glues and adhesives, printers, and photocopy machines also give off dangerous gases, including formaldehyde. Like cigarette smoke, formaldehyde is a carcinogen, which means it can cause cancer.

Carbon Monoxide Carbon monoxide (CO) is a poisonous gas that is produced whenever charcoal, natural gas, kerosene, or other fuels are burned. CO poisoning can cause serious illness or death. Fuel-burning stoves and heaters must be designed to prevent CO from building up indoors. CO is colorless and odorless, so it is difficult to detect. Alarms that provide warning of a dangerous buildup of CO are being used in more and more homes.

Radon Radon is a naturally occurring, radioactive gas that is given off by some types of rock and soil, as shown in **Figure 15.** Radon has no color or odor. It can seep into basements and the lower floors of buildings. Radon exposure is the second leading cause of lung cancer in this country. A radon detector sounds an alarm when levels of the gas in indoor air become too high. If radon is present, increasing a building's ventilation can eliminate any damaging effects.

Figure 15 The map shows the potential for radon exposure in different parts of the United States. **Identify** *the area of the country with soils that produce the most radon gas.*

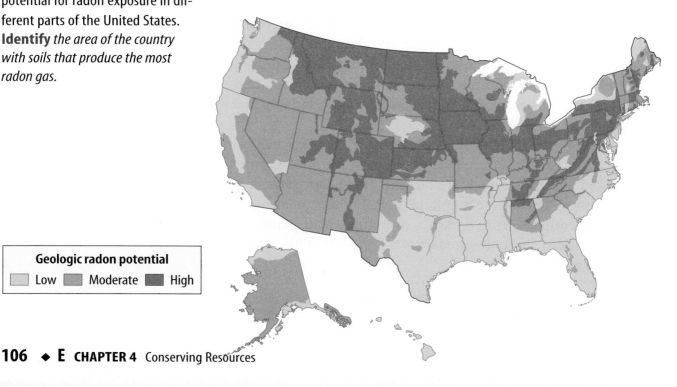

Geologic radon potential
Low Moderate High

When rain falls on roads and parking lots, it can wash oil and grease onto the soil and into nearby streams.

Rain can wash agricultural pesticides and fertilizers into lakes, streams, or oceans.

Industrial wastes are sometimes released directly into surface waters.

Figure 16 Pollution of surface waters can occur in several ways, as shown above.

Water Pollution

Pollutants enter water, too. Air pollutants can drift into water or be washed out of the sky by rain. Rain can wash land pollutants into waterways, as shown in **Figure 16.** Wastewater from factories and sewage-treatment plants often is released into waterways. In the United States and many other countries, laws require that wastewater be treated to remove pollutants before it is released. But, in many parts of the world, wastewater treatment is not always possible. Pollution also enters water when people dump litter or waste materials into rivers, lakes, and oceans.

Surface Water Some water pollutants poison fish and other wildlife, and can be harmful to people who swim in or drink the water. For example, chemical pesticides sprayed on farmland can wash into lakes and streams. These chemicals can harm the insects that fish, turtles, or frogs rely on for food. Shortages of food can lead to deaths among water-dwelling animals. Some pollutants, especially those containing mercury and other metals, can build up in the tissues of fish. Eating contaminated fish and shellfish can transfer these metals to people, birds, and other animals. In some areas, people are advised not to eat fish or shellfish taken from polluted waterways.

Algal blooms are another water pollution problem. Raw sewage and excess fertilizer contain large amounts of nitrogen. If they are washed into a lake or pond, they can cause the rapid growth of algae. When the algae die, they are decomposed by huge numbers of bacteria that use up much of the oxygen in the water. Fish and other organisms can die from a lack of oxygen in the water.

Figure 17 In 1996, the oil tanker *Sea Empress* spilled more than 72 million kg of oil into the sea along the coast of Wales. More than $40 million was spent on the cleanup effort, but thousands of ocean organisms were destroyed, including birds, fish, and shellfish.

Ocean Water Rivers and streams eventually flow into oceans, bringing their pollutants along. Also, polluted water can enter the ocean in coastal areas where factories, sewage-treatment plants, or shipping activities are located. Oil spills are a well-known ocean pollution problem. About 4 billion kg of oil are spilled into ocean waters every year. Much of that oil comes from ships that use ocean water to wash out their fuel tanks. Oil also can come from oil tanker wrecks, as shown in **Figure 17.**

Groundwater Pollution can affect water that seeps underground, as shown in **Figure 18.** Groundwater is water that collects between particles of soil and rock. It comes from precipitation and runoff that soaks into the soil. This water can flow slowly through permeable layers of rock called aquifers. If this water comes into contact with pollutants as it moves through the soil and into an aquifer, the aquifer could become polluted. Polluted groundwater is difficult—and sometimes impossible—to clean. In some parts of the country, chemicals leaking from underground storage tanks have created groundwater pollution problems.

Figure 18 Water from rainfall slowly filters through sand or soil until it is trapped in underground aquifers. Pollutants picked up by the water as it filters through the soil can contaminate water wells.

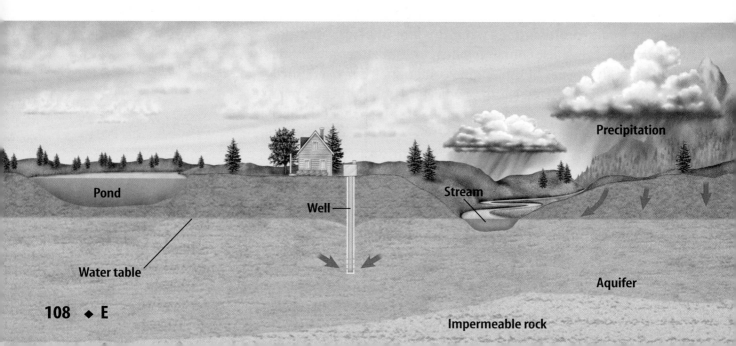

Pond

Well

Stream

Precipitation

Water table

Aquifer

Impermeable rock

Contour plowing reduces the downhill flow of water.

Soil Loss

Fertile topsoil is important to plant growth. New topsoil takes hundreds or thousands of years to form. The Launch Lab at the beginning of this chapter shows that rain washes away loose topsoil. Wind also blows it away. The movement of soil from one place to another is called **erosion** (ih ROH zhun). Eroded soil that washes into a river or stream can block sunlight and slow photosynthesis. It also can harm fish, clams, and other organisms. Erosion is a natural process, but human activities increase it. When a farmer plows a field or a forest is cut down, soil is left bare. Bare soil is more easily carried away by rain and wind. **Figure 19** shows some methods farmers use to reduce soil erosion.

Soil Pollution

Soil can become polluted when air pollutants drift to the ground or when water leaves pollutants behind as it flows through the soil. Soil also can be polluted when people toss litter on the ground or dispose of trash in landfills.

Solid Wastes What happens to the trash you throw out every week? What do people do with old refrigerators, TVs, and toys? Most of this solid waste is dumped in landfills. Most landfills are designed to seal out air and water. This helps prevent pollutants from seeping into surrounding soil, but it slows normal decay processes. Even food scraps and paper, which usually break down quickly, can last for decades in a landfill. In populated areas, landfills fill up quickly. Reducing the amount of trash people generate can reduce the need for new landfills.

Figure 19 The farming methods shown here help prevent soil erosion.
Infer *why soil erosion is a concern for farmers.*

On steep hillsides, flat areas called terraces reduce downhill flow.

In strip cropping, cover crops are planted between rows to reduce wind erosion.

In no-till farming, soil is never left bare.

Figure 20 Leftover paints, batteries, pesticides, drain cleaners, and medicines are hazardous wastes that should not be discarded in the trash. They should never be poured down a drain, onto the ground, or into a storm sewer. Most communities have collection facilities where people can dispose of hazardous materials like these.

Hazardous Wastes Waste materials that are harmful to human health or poisonous to living organisms are **hazardous wastes.** They include dangerous chemicals, such as pesticides, oil, and petroleum-based solvents used in industry. They also include radioactive wastes from nuclear power plants, from hospitals that use radioactive materials to treat disease, and from nuclear weapons production. Many household items also are considered hazardous, such as those shown in **Figure 20.** If these materials are dumped into landfills, they could seep into the soil, surface water, or groundwater over time. Hazardous wastes usually are handled separately from trash. They are treated in ways that prevent environmental pollution.

✔ **Reading Check** *What are hazardous wastes?*

section ② review

Summary

Air Pollution and Acid Precipitation
- Vehicles, volcanoes, forest fires, and even wind-blown dust and sand can cause air pollution.
- Acid rain washes nutrients from the soil, which can harm plants.

Greenhouse Effect and Ozone Depletion
- CO_2 is a greenhouse gas that helps warm Earth.
- The ozone layer protects life on Earth.

Indoor Air Pollution, Water Pollution, Soil Loss, and Soil Pollution
- Pollutants can build up inside of buildings.
- There are many sources of water pollutants.
- Wind and rain can erode bare soil.
- Pollutants in soil decay more slowly than in air.

Self Check

1. **List** four ways that air pollution affects the environment.
2. **Explain** how an algal bloom can affect other pond organisms.
3. **Describe** possible causes and effects of ozone depletion.
4. **Think Critically** How could hazardous wastes in landfills eventually affect groundwater?

Applying Math

5. **Solve a One-Step Equation** A solution of pH 4 is 10 times more acidic than one of pH 5, and it is 10 times more acidic than a solution of pH 6. How many times more acidic is the solution of pH 4 than the one of pH 6?

Science Online booke.msscience.com/self_check_quiz

The Greenhouse Effect

You can create models of Earth with and without heat-reflecting green-house gases. Then, experiment with the models to observe the greenhouse effect.

▶ Real-World Question

How does the greenhouse effect influence temperatures on Earth?

Goals
- **Observe** the greenhouse effect.
- **Describe** the effect that a heat source has on an environment.

Materials
1-L clear-plastic, soft-drink bottles with tops cut off and labels removed (2)
thermometers (2)
*temperature probe
potting soil
masking tape
plastic wrap
rubber band
lamp with 100-W lightbulb
watch or clock with second hand
*Alternate materials

Safety Precautions 🥽 🧤 🖐️ 🧹 📋

▶ Procedure

1. Copy the data table and use it to record your temperature measurements.
2. Put an equal volume of potting soil in the bottom of each container.
3. Use masking tape to attach a thermometer to the inside of each container. Place each thermometer at the same height above the soil. Shield each thermometer bulb by putting a double layer of masking tape over it.

Changes in Temperature

Time (min)	Open Container Temperature (°C)	Closed Container Temperature (°C)
0		
2	Do not write in this book.	
4		
6		

4. Seal the top of one container with plastic wrap held in place with a rubber band.
5. Place the lamp with the exposed 100-W lightbulb between the two containers and exactly 1 cm away from each. Do not turn on the light.
6. Let the setup sit for 5 min, then record the temperature in each container.
7. Turn on the light. Record the temperature in each container every 2 min for 15 min to 20 min. Graph the results.

▶ Conclude and Apply

1. **Compare and contrast** temperatures in each container at the end of the experiment.
2. **Infer** What does the lightbulb represent in this experimental model? What does the plastic wrap represent?

Communicating Your Data

Average the data obtained in the experiments conducted by all the groups in your class. Prepare a line graph of these data. **For more help, refer to the** Science Skill Handbook.

The Three Rs of Conservation

What You'll Learn

- **Recognize** ways you can reduce your use of natural resources.
- **Explain** how you can reuse resources to promote conservation.
- **Describe** how many materials can be recycled.

Why It's Important

Conservation preserves resources and reduces pollution.

Review Vocabulary
reprocessing: to subject to a special process or treatment in preparation for reuse

New Vocabulary
- recycling

Figure 21 Worn-out automobile tires can have other useful purposes.

Conservation

A teacher travels to school in a car pool. In the school cafeteria, students place glass bottles and cans in separate containers from the rest of the garbage. Conservation efforts like these can help prevent shortages of natural resources, slow growth of landfills, reduce pollution levels, and save people money. Every time a new landfill is created, an ecosystem is disturbed. Reducing the need for landfills is a major benefit of conservation. The three Rs of conservation are reduce, reuse, and recycle.

Reduce

You contribute to conservation whenever you reduce your use of natural resources. You use less fossil fuel when you walk or ride a bicycle instead of taking the bus or riding in a car. If you buy a carton of milk, reduce your use of petroleum by telling the clerk you don't need a plastic bag to carry it in.

You also can avoid buying things you don't need. For example, most of the paper, plastic, and cardboard used to package items for display on store shelves is thrown away as soon as the product is brought home. You can look for products with less packaging or with packaging made from recycled materials. What are some other ways you can reduce your use of natural resources?

Reuse

Another way to help conserve natural resources is to use items more than once. Reusing an item means using it again without changing it or reprocessing it, as shown in **Figure 21.** Bring reusable canvas bags to the grocery store to carry home your purchases. Donate clothes you've outgrown to charity so that others can reuse them. Take reusable plates and utensils on picnics instead of disposable paper items.

Recycle

If you can't avoid using an item, and if you can't reuse it, the next best thing is to recycle it. **Recycling** is a form of reuse that requires changing or reprocessing an item or natural resource. If your city or town has a curbside recycling program, you already separate recyclables from the rest of your garbage. Materials that can be recycled include glass, metals, paper, plastics, and yard and kitchen waste.

Reading Check *How is recycling different from reusing?*

Plastics Plastic is more difficult to recycle than other materials, mainly because several types of plastic are in use. A recycle code marked on every plastic container indicates the type of plastic it is made of. Plastic soft-drink bottles, like the one shown in **Figure 22,** are made of type 1 plastic and are the easiest to recycle. Most plastic bags are made of type 2 or type 4 plastic; they can be reused as well as recycled. Types 6 and 7 can't be recycled at all because they are made of a mixture of different plastics. Each type of plastic must be separated carefully before it is recycled because a single piece of a different type of plastic can ruin an entire batch.

Figure 22 Many soft-drink bottles are made of PETE, which is the most common type of recyclable plastic. It can be melted down and spun into fibers to make carpets, paintbrushes, rope, and clothing. **Identify** *other products made out of recycled materials.*

Metals The manufacturing industry has been recycling all kinds of metals, especially steel, for decades. At least 25 percent of the steel in cans, appliances, and automobiles is recycled steel. Up to 100 percent of the steel in plates and beams used to build skyscrapers is made from reprocessed steel. About one metric ton of recycled steel saves about 1.1 metric tons of iron ore and 0.5 metric ton of coal. Using recycled steel to make new steel products reduces energy use by 75 percent. Other metals, including iron, copper, aluminum, and lead also can be recycled.

You can conserve metals by recycling food cans, which are mostly steel, and aluminum cans. It takes less energy to make a can from recycled aluminum than from raw materials. Also, remember that recycled cans do not take up space in landfills.

Glass When sterilized, glass bottles and jars can be reused. They also can be melted and re-formed into new bottles, especially those made of clear glass. Most glass bottles already contain at least 25 percent recycled glass. Glass can be recycled again and again. It never needs to be thrown away. Recycling about one metric ton of glass saves more than one metric ton of mineral resources and reduces the energy used to make new glass by 25 percent or more.

Applying Science

What items are you recycling at home?

Many communities have recycling programs. Recyclable items may be picked up at the curbside, taken to a collection site, or the resident may hire a licensed recycling handler to pick them up. What do you recycle in your home?

Identifying the Problem

This bar graph shows the recycling rates in the U. S. of six types of household items for the years 1990, 1995, and 2000. What are you and your classmates' recycling rates?

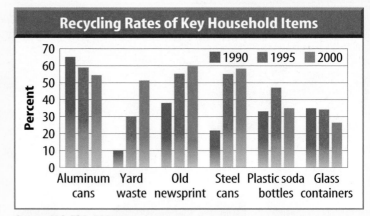

Source: U.S. EPA, 2003

Solving the Problem

For one week, list each glass, plastic, and aluminum item you use. Note which items you throw away and which ones you recycle. Calculate the percentage of glass, plastic, and aluminum you recycled. How do your percentages compare with those on the graph?

Paper Used paper is recycled into paper towels, insulation, newsprint, cardboard, and stationery. Ranchers and dairy farmers sometimes use shredded paper instead of straw for bedding in barns and stables. Used paper can be made into compost. Recycling about one metric ton of paper saves 17 trees, more than 26,000 L of water, close to 1,900 L of oil, and more than 4,000 kW of electric energy. You can do your part by recycling newspapers, notebook and printer paper, cardboard, and junk mail.

Reading Check *What nonrenewable resource(s) do you conserve by recycling paper?*

Figure 23 Composting is a way of turning plant material you would otherwise throw away into rich garden soil. Dry leaves and weeds, grass clippings, vegetable trimmings, and nonmeat food scraps can be composted.

Compost Grass clippings, leaves, and fruit and vegetable scraps that are discarded in a landfill can remain there for decades without breaking down. The same items can be turned into soil-enriching compost in just a few weeks, as shown in **Figure 23.** Many communities distribute compost bins to encourage residents to recycle fruit and vegetable scraps and yard waste.

Buy Recycled People have become so good at recycling that recyclable materials are piling up, just waiting to be put to use. You can help by reading labels when you shop and choosing products that contain recycled materials. What other ways of recycling natural resources can you think of?

section 3 review

Summary

Conservation
- The three Rs of conservation are reduce, reuse, and recycle.

Reduce
- You can contribute to conservation by reducing your use of natural resources.

Reuse
- Some items can be used more than once, such as reusable canvas bags for groceries.

Recycle
- Some items can be recycled, including some plastics, metal, glass, and paper.
- Grass clippings, leaves, and fruit and vegetable scraps can be composted into rich garden soil.

Self Check

1. **Describe** at least three actions you could take to reduce your use of natural resources.
2. **Describe** how you could reuse three items people usually throw away.
3. **Think Critically** Why is reusing something better than recycling it?

Applying Skills

4. **Make and Use Tables** Make a table of data of the number of aluminum cans thrown away in the United States: 2.7 billion in 1970; 11.1 billion in 1974; 21.3 billion in 1978; 22.7 billion in 1982; 35.0 billion in 1986; 33.8 billion in 1990; 38.5 billion in 1994; 45.5 billion in 1998; 50.7 billion in 2001.

Model and Invent

Solar Cooking

Real-World Question

The disappearance of forests in some places on Earth has made firewood extremely difficult and expensive to obtain. People living in these regions often have to travel long distances or sell some of their food to get firewood. This can be a serious problem for people who may not have much food to begin with. Is there a way they could cook food without using firewood? How would you design and build a cooking device that uses the Sun's energy?

Goals

- **Research** designs for solar panel cookers or box cookers.
- **Design** a solar cooker that can be used to cook food.
- **Plan** an experiment to measure the effectiveness of your solar cooker.

Possible Materials

poster board
cardboard boxes
aluminum foil
string
wire coat hangers
clear plastic sheets
*oven bags
black cookware
thermometer
stopwatch
*timer
glue
tape
scissors
*Alternate materials

Safety Precautions

WARNING: *Be careful when cutting materials. Your solar cooker will get hot. Use insulated gloves or tongs to handle hot objects.*

Make the Model

1. **Design** a solar cooker. In your Science Journal, explain why you chose this design and draw a picture of it.

2. **Write** a summary explaining how you will measure the effectiveness of your solar cooker. What will you measure? How will you collect and organize your data? How will you present your results?

3. **Compare** your solar cooker design to those of other students.

4. Share your experimental plan with students in your class. Discuss the reasoning behind your plan. Be specific about what you intend to test and how you are going to test it.

5. Make sure your teacher approves your plan before you start working on your model.

6. Using all of the information you have gathered, construct a solar cooker that follows your design.

◉ *Test the Model*

1. **Test** your design to determine how well it works. Try out a classmate's design. How do the two compare?

◉ *Analyze Your Data*

1. Combine the results for your entire class and decide which type of solar cooker was most effective. How could you design a more effective solar cooker, based on what you learned from this activity?

2. **Infer** Do you think your results might have been different if you tested your solar cooker on a different day? Explain. Why might a solar cooker be more useful in some regions of the world than in others?

◉ *Conclude and Apply*

1. **Infer** Based on what you've read and the results obtained by you and your classmates, do you think that your solar cooker could boil water? Explain.

2. **Compare** the amount of time needed to cook food with a solar cooker and with more traditional cooking methods. Assuming plenty of sunlight is available, would you prefer to use a solar cooker or a traditional oven? Explain.

Prepare a demonstration showing how to use a solar cooker. Present your demonstration to another class of students or to a group of friends or relatives. **For more help, refer to the** Science Skill Handbook.

Beauty Plagiarized
by Amitabha Mukerjee

I wandered lonely as a cloud –
Except for a motorboat,
Nary a soul in sight.
Beside the lake beneath the trees,
Next to the barbed wire fence,
There was a picnic table
And beer bottle caps from many years.
A boat ramp to the left,
And the chimney from a power station on the
    other side,
A summer haze hung in the air,
And the lazy drone of traffic far away.

Crimson autumn of mists and mellow fruitfulness
Blue plastic covers the swimming pools
The leaves fall so I can see
Dark glass reflections in the building
That came up
where the pine cones crunched underfoot . . .

And then it is snow
White lining on trees and rooftops . . .
And through my windshield wipers
The snow is piled dark and grey . . .
Next to my driveway where I check my mail
Little footprints on fresh snow —
A visiting rabbit.

I knew a bank where the wild thyme blew
Over-canopied with luscious woodbine
It is now a landfill —
Fermentation of civilization
Flowers on TV
Hyacinth rose tulip chrysanthemum
Acres of colour
Wind up wrapped in decorous plastic,
In this landfill where oxlips grew. . .

Understanding Literature

Cause and Effect Recognizing cause-and-effect relationships can help you make sense out of what you read. One event causes another event. The second event is the effect of the first event. In the poem, the author describes the causes and effects of pollution and waste. What effects do pollution and the use of nonrenewable resources have on nature in the poem?

Respond to the Reading

1. To plagiarize is to copy without giving credit to the source. In this poem, who or what has plagiarized beauty?
2. What do the four verses in the poem correspond to?
3. **Linking Science and Writing** Write a poem that shows how conservation methods could restore the beauty in nature.

 The poet makes a connection between the four seasons of the year and the pollution and waste products created by human activity, or civilization. For example, in the spring, a landfill for dumping garbage replaces a field of wildflowers. Describing four seasons instead of one reinforces the poet's message that the beauty of nature has been stolen, or plagiarized.

Reviewing Main Ideas

Section 1 Resources

1. Natural resources are the parts of the environment that supply materials needed for the survival of living organisms.

2. Renewable resources are being replaced continually by natural processes.

3. Nonrenewable resources cannot be replaced or are replaced very slowly.

4. Energy sources include fossil fuels, wind, solar energy, geothermal energy, hydroelectric power, and nuclear power.

Section 2 Pollution

1. Most air pollution is made up of waste products from the burning of fossil fuels.

2. The greenhouse effect is the warming of Earth by a blanket of heat-reflecting gases in the atmosphere.

3. Water can be polluted by acid rain and by the spilling of oil or other wastes into waterways.

4. Solid wastes and hazardous wastes dumped on land or disposed of in landfills can pollute the soil. Erosion can cause the loss of fertile topsoil.

Section 3 The Three Rs of Conservation

1. You can reduce your use of natural resources in many ways.

2. Reusing items is an excellent way to practice conservation.

3. In recycling, materials are changed in some way so that they can be used again.

4. Materials that can be recycled include paper, metals, glass, plastics, yard waste, and nonmeat kitchen scraps.

Visualizing Main Ideas

Copy and complete the following concept map using the terms smog, acid precipitation, *and* ozone depletion.

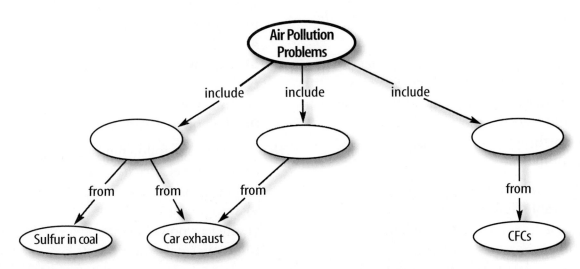

Using Vocabulary

acid precipitation p. 103	nonrenewable
erosion p. 109	resource p. 95
fossil fuel p. 96	nuclear energy p. 98
geothermal energy p. 99	ozone depletion p. 105
greenhouse effect p. 104	petroleum p. 95
hazardous waste p. 110	pollutant p. 102
hydroelectric power p. 97	recycling p. 113
natural resource p. 94	renewable resource p. 94

Explain the differences in the vocabulary words given below. Then explain how the words are related. Use complete sentences in your answers.

1. fossil fuel—petroleum

2. erosion—pollutant

3. ozone depletion—acid precipitation

4. greenhouse effect—fossil fuels

5. hazardous wastes—nuclear energy

6. hydroelectric power—fossil fuels

7. acid precipitation—fossil fuels

8. ozone depletion—pollutant

9. recycle—nonrenewable resources

10. geothermal energy—fossil fuels

Checking Concepts

Choose the word or phrase that best answers the question.

11. An architect wants to design a solar house in the northern hemisphere. For maximum warmth, which side of the house should have the most windows?
    - **A)** north
    - **B)** south
    - **C)** east
    - **D)** west

12. Of the following, which is considered a renewable resource?
    - **A)** coal
    - **B)** oil
    - **C)** sunlight
    - **D)** aluminum

Use the photo below to answer question 13.

13. Which energy resource is shown in the photo?
    - **A)** solar energy
    - **B)** geothermal energy
    - **C)** hydroelectric energy
    - **D)** photovoltaic energy

14. Which of the following is a fossil fuel?
    - **A)** wood
    - **B)** oil
    - **C)** nuclear power
    - **D)** photovoltaic cell

15. Which of the following contributes to ozone depletion?
    - **A)** carbon dioxide
    - **B)** radon
    - **C)** CFCs
    - **D)** carbon monoxide

16. What is a substance that contaminates the environment called?
    - **A)** acid rain
    - **B)** pollution
    - **C)** pollutant
    - **D)** ozone

17. If there were no greenhouse effect in Earth's atmosphere, which of the following statements would be true?
    - **A)** Earth would be much hotter.
    - **B)** Earth would be much colder.
    - **C)** The temperature of Earth would be the same.
    - **D)** The polar ice caps would melt.

18. Which of the following can change solar energy into electricity?
    - **A)** photovoltaic cells
    - **B)** smog
    - **C)** nuclear power plants
    - **D)** geothermal power plants

Thinking Critically

19. **Explain** how geothermal energy is used to produce electricity.

20. **Infer** why burning wood and burning fossil fuels produce similar pollutants.

Use the photos below to answer question 21.

21. **Draw a Conclusion** Which would make a better location for a solar power plant—a polar region (left) or a desert region (right)? Why?

22. **Explain** why it is beneficial to grow a different crop on soil after the major crop has been harvested.

23. **Infer** Is garbage a renewable resource? Why or why not?

24. **Summarize** Solar, nuclear, wind, water, and geothermal energy are alternatives to fossil fuels. Are they all renewable? Why or why not?

25. **Draw Conclusions** Would you save more energy by recycling or reusing a plastic bag?

26. **Recognize Cause and Effect** Forests use large amounts of carbon dioxide during photosynthesis. How might cutting down a large percentage of Earth's forests affect the greenhouse effect?

27. **Form a hypothesis** about why Americans throw away more aluminum cans each year.

28. **Compare and contrast** contour farming, terracing, strip cropping, and no-till farming.

Performance Activities

29. **Poster** Create a poster to illustrate and describe three things students at your school can do to conserve natural resources.

Applying Math

Use the table below to answer questions 30 and 31.

Estimated Recycling Rates	
Item	**Percent Recycled**
Aluminum cans	60
Glass beverage bottles	31
Plastic soft-drink containers	37
Newsprint	56
Magazines	23

30. **Recycling Rates** Make a bar graph of the data above.

31. **Bottle Recycling** For every 1,000 glass beverage bottles that are produced, how many are recycled?

32. **Nonrenewable Resources** 45.8 billion (45,800,000,000) cans were thrown away in 2000. If it takes 33.79 cans to equal one pound and the average scrap value is \$0.58/lb, then what was the total dollar value of the discarded cans?

33. **Ozone Depletion** The thin ozone layer called the "ozone hole" over Antarctica reached nearly 27,000,000 km^2 in 1998. To conceptualize this, the United States has a geographical area of 9,363,130 km^2. How much larger is the "ozone hole" in comparison to the United States?

34. **Increased CO_2 Levels** To determine the effects of increased CO_2 levels in the atmosphere, scientists increased the CO_2 concentration by 70 percent in an enclosed rain forest environment. If the initial CO_2 concentration was 430 parts per million, what was it after the increase?

Part 1 | Multiple Choice

Record your answers on the answer sheet provided by your teacher or on a sheet of paper.

1. From what natural resource are plastics, paints, and gasoline made?
   A. coal
   B. petroleum
   C. iron ore
   D. natural gas

Use the illustration below to answer questions 2–4.

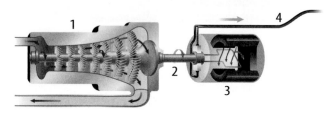

2. What is produced by the mechanism shown in the illustration?
   A. electricity
   B. coal
   C. petroleum
   D. plastic

3. In which section are the turbine blades found?
   A. 1
   B. 2
   C. 3
   D. 4

4. Which section represents the generator?
   A. 1
   B. 2
   C. 3
   D. 4

5. Which of the following is necessary for the production of hydroelectric power?
   A. wind
   B. access to a river
   C. exposure to sunlight
   D. heat from below Earth's crust

6. With which type of alternative energy are photovoltaic cells used?
   A. hydroelectric power
   B. geothermal energy
   C. nuclear energy
   D. solar energy

7. Which of the following is a type of air pollution that results when sunlight reacts with pollutants produced by burning fuels?
   A. ozones
   B. acid rain
   C. smog
   D. UV radiation

Use the photograph below to answer questions 8 and 9.

8. What is the name of the method of farming illustrated above?
   A. contour plowing
   B. strip cropping
   C. terracing
   D. no-till farming

9. What is the purpose of the method shown in the illustration?
   A. to decrease soil erosion from wind
   B. to decrease soil erosion from water flow
   C. to decrease acid rain production
   D. to increase the return of nutrients to the soil

Test-Taking Tip

Qualifying Terms Look for qualifiers in a question. Such questions are not looking for absolute answers. Qualifiers could be words such as most likely, most common, or least common.

Question 18 The qualifier in this question is *possible*. This indicates that there is uncertainty about the effects of global warming.

Part 2 | Short Response/Grid In

Record your answers on the answer sheet provided by your teacher or on a sheet of paper.

10. Give one example of a renewable source of energy and one example of a nonrenewable source of energy.

Use the illustration below to answer questions 11 and 12.

11. What type of alternative energy is the girl using in the diagram?

12. Name one benefit and one drawback to using this type of energy for cooking.

13. What are two ways that smog can be reduced?

14. A group of students collects rain outside their classroom, then tests the pH of the collected rain. The pH of the rain is 7.2. Can the students say that their rain is acid rain? Why or why not?

15. Why do we depend on the greenhouse effect for survival?

16. What is the cause of algal blooms in lakes and ponds?

Part 3 | Open Ended

Record your answers on a sheet of paper.

17. Are renewable resources always readily available? Explain.

18. What are the possible worldwide effects of global warming? What causes global warming? Why do some people think that using fossil fuels less will decrease global warming?

19. A family lives in a house that uses solar panels to heat the hot water, a wood-burning stove to heat the house, and a windmill for pumping water from a well into a tower where it is stored and then piped into the house as needed. What would be the result if there was no sunlight for two weeks?

Use the illustration below to answer question 20.

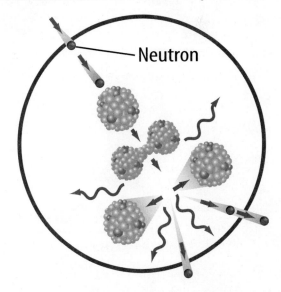

Neutron

20. What does the diagram represent?

21. Explain how different kinds of plastics are recycled.

The BIG Idea

Biodiversity is important for all organisms on Earth.

SECTION 1
Biodiversity
Main Idea Biodiversity is reduced by pollution, habitat loss, and extinction.

SECTION 2
Conservation Biology
Main Idea Habitat restoration and preservation are two ways to conserve biodiversity.

Conserving Life

A Band Helps Preserve Biodiversity

Before formulating a plan to preserve a species, a conservation biologist must first understand what it is that the species needs. Biologists employ a number of techniques to study species, including marking individuals.

Science Journal List an organism that you feel is important for maintaining biodiversity and explain why it's important.

Start-Up Activities

Recognize Environmental Differences

Do human actions affect the number of species present in an ecosystem? What happens to organisms that are pushed out of an area when the environment changes?

1. Using string or tape, mark off a 1-m × 1-m area of lawn, sports field, or sidewalk.

2. Count and record the different species of plants and animals present in the sample area. Don't forget to include insects or birds that fly over the area, or organisms found by gently probing into the soil.

3. Repeat steps 1 and 2 in a partially wooded area, in a weedy area, or at the edge of a pond or stream.

4. **Think Critically** Make notes about your observations in your Science Journal. What kinds of human actions could have affected the environment of each sample area?

Science Online **Preview this chapter's content and activities at** booke.msscience.com

Biodiversity Make the following Foldable to help you identify what you already know, what you want to know, and what you learned about biodiversity.

STEP 1 Fold a vertical sheet of paper from side to side. Make the front edge about 1.25 cm shorter than the back edge.

STEP 2 **Turn** lengthwise and **fold** into thirds.

STEP 3 **Unfold and cut** only the top layer along both folds to make three tabs. **Label** each tab as shown.

Ask Questions Before you read the chapter, write what you already know about biodiversity under the left tab of your Foldable, and write questions about what you'd like to know under the center tab. After you read the chapter, list what you learned under the right tab.

Identify the Main Idea

① Learn It! Main ideas are the most important ideas in a paragraph, section, or chapter. Supporting details are facts or examples that explain the main idea. Understanding the main idea allows you to grasp the whole picture.

② Picture It! Read the following paragraph. Draw a graphic organizer like the one below to show the main idea and supporting details.

> Carbon dioxide gas (CO_2) is released into the atmosphere when wood, coal, gas, or any other fuel is burned. People burn large amounts of fuel, and this is contributing to an increase in the percentage of CO_2 in the atmosphere. An increase in CO_2 could raise Earth's temperature by a few degrees. This average temperature rise, called global warming, might lead to climatic changes that could affect biodiversity. For example, portions of the polar ice caps could melt, causing floods in coastal ecosystems around the world.
>
> —*from page 136*

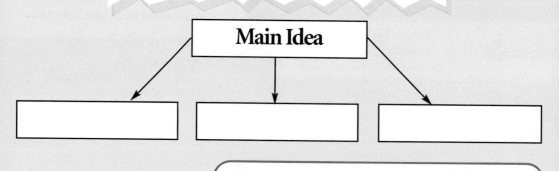

③ Apply It! Pick a paragraph from another section of this chapter and diagram the main ideas as you did above.

Reading Tip

The main idea is often the first sentence in a paragraph but not always.

Target Your Reading

Use this to focus on the main ideas as you read the chapter.

1 Before you read the chapter, respond to the statements below on your worksheet or on a numbered sheet of paper.

- Write an **A** if you **agree** with the statement.
- Write a **D** if you **disagree** with the statement.

2 After you read the chapter, look back to this page to see if you've changed your mind about any of the statements.

- If any of your answers changed, explain why.
- Change any false statements into true statements.
- Use your revised statements as a study guide.

Science Online

Print out a worksheet of this page at booke.msscience.com

Before You Read A or D		Statement	After You Read A or D
	1	Biodiversity is the number of organisms in an ecosystem.	
	2	Extinction is a normal part of nature.	
	3	A species in danger of extinction is labeled as threatened.	
	4	Small areas of habitat are normally more diverse than larger areas.	
	5	Ozone depletion occurs over much of Earth.	
	6	There are laws to protect species and their habitats.	
	7	Habitats that have been harmed can be restored.	
	8	Preservation and restoration of habitats automatically protect species living there.	
	9	Individual organisms living in captivity can never be released back into the wild.	

Biodiversity

as you read

What You'll Learn

- **Define** biodiversity.
- **Explain** why biodiversity is important in an ecosystem.
- **Identify** factors that limit biodiversity in an ecosystem.

Why It's Important

Knowledge of biological diversity can lead to strategies for preventing the loss of species.

⊘ Review Vocabulary

mammal: endothermic verte-brate with mammary glands and hair growing from their skin

New Vocabulary

- biodiversity
- extinct species
- endangered species
- threatened species
- introduced species
- native species
- acid rain
- ozone depletion

The Variety of Life

Imagine walking through a forest ecosystem like the one shown in **Figure 1.** Trees, shrubs, and small plants are every-where. You see and hear squirrels, birds, and insects. You might notice a snake or mushrooms. Hundreds of species live in this forest. Now, imagine walking through a wheat field. You see only a few species—wheat plants, insects, and weeds. The forest con-tains more species than the wheat field does. The forest has a higher biological diversity, or biodiversity. **Biodiversity** refers to the variety of life in an ecosystem.

Measuring Biodiversity The common measure of biodiver-sity is the number of species that live in an area. For example, a coral reef can be home to thousands of species including corals, fish, algae, sponges, crabs, and worms. A coral reef has greater biodiversity than the shallow waters that surround it. Before deep-sea exploration, scientists thought that few organisms could live in dark, deep-sea waters. Although the number of organisms living there is likely to be less than the number of organisms on a coral reef, we know that the species biodiversity of deep-sea waters is as great as that of a coral reef.

Figure 1 The forest has more species and is richer in biodiversity than a wheat field.

Figure 2 This map shows the number of mammal species found in three North American countries. In general, biodiversity increases as you get closer to the equator.

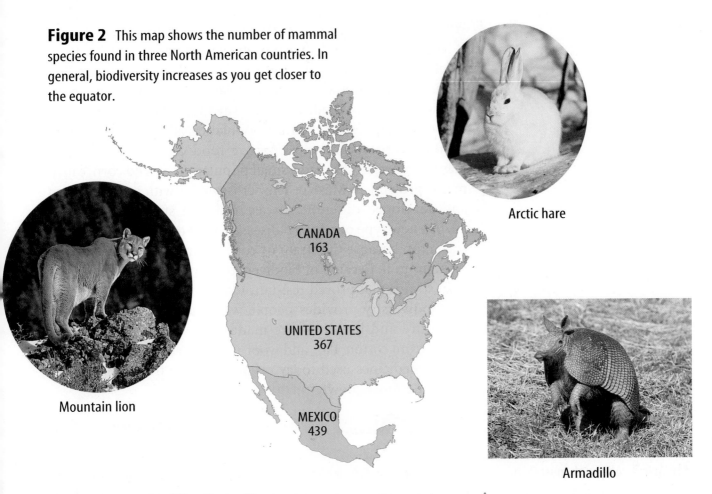

Arctic hare

Mountain lion

CANADA
163

UNITED STATES
367

MEXICO
439

Armadillo

Differences in Biodiversity Biodiversity tends to increase as you move toward the equator because temperatures tend to be warmer. For example, Costa Rica is a Central American country about the size of West Virginia. Yet it is home to as many bird species as there are in the United States and Canada combined. **Figure 2** compares mammal biodiversity in three North American countries. Ecosystems with the highest biodiversity usually have warm, moist climates. In fact, tropical regions contain two-thirds of all of Earth's land species.

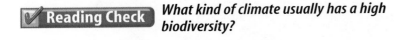 **Reading Check** *What kind of climate usually has a high biodiversity?*

Why is biodiversity important?

Would you rather have a summer picnic on a paved parking lot or on cool grass under a shade tree near a stream? Many people find pleasure in nature's biodiversity. Perhaps you like to visit scenic areas to hike, swim, relax, or enjoy the views. Many painters, writers, and musicians find inspiration for their work by spending time outdoors. However, beauty and pleasure are not the only reasons why biodiversity is important.

Humans Need Biodiversity What foods do you like to eat? If you eat meat or fish, your meals probably include beef, chicken, pork, tuna, shrimp, or clams. Rice, peanuts, green beans, strawberries, or carrots are plant foods you might enjoy. Eating a variety of foods is a good way to stay healthy. Hundreds of species help feed the human population all around the world.

Biodiversity can help improve food crops. For example, plant breeders discovered a small population of wild corn, called maize, growing in Mexico. They crossbred it with domestic corn and developed different, disease-resistant strains of corn. In 1970, a new type of a fungal disease wiped out much of the U.S. corn crop because the strain of corn commonly grown was not resistant to it. Because of biodiversity, farmers could plant other corn strains that proved resistant to this new disease.

Biodiversity provides people with many useful materials. Furniture and buildings are made from wood and bamboo. Fibers from cotton, flax, and wool are woven into clothing. Most of the medicines used today originally came from wild plants, including those shown in **Figure 3.**

Reading Check *Why is biodiversity important to people?*

Figure 3 Although most medicines are made in factories, many originally came from wild plants.

Maintaining Stability Forests usually contain many different kinds of plants. If one type of plant disappears, the forest still exists. Biodiversity allows for stability in an ecosystem.

An extract from the bark of the Pacific yew was traditionally used to treat arthritis and rheumatism. It is now known that the bark contains taxol, which is a treatment for some types of cancer.

The velvet bean plant is the original source of L-dopa, a drug used to treat Parkinson's disease.

The bark of the cinchona tree contains quinine, which has been used to cure malaria in millions of people.

A loss of biodiversity can weaken an ecosystem. In a vineyard, as shown in **Figure 4,** vines grow close together. If a disease infects one grapevine, it could move easily from one plant to the next. Soon, the entire vineyard could be infected. Many farmers and gardeners have found that planting alternate rows of different crops can help prevent disease and reduce or eliminate the need for pesticides.

Figure 4 Disease has spread from one grapevine to another in this vineyard.

Applying Math · Solve a One-Step Equation

MEASURING BIODIVERSITY Temperate rain forest ecosystems in Alaska and California are similar. Which one has the higher mammal and bird biodiversity?

Biodiversity

Number of Species	California	Alaska
Mammals	49	38
Birds	94	79

Solution

1 *This is what you know:*

49 = mammal species in California

94 = bird species in California

38 = mammal species in Alaska

79 = bird species in Alaska

2 *This is what you need to find out:*

Which of these ecosystems has the greater bird and mammal biodiversity?

3 *This is the procedure you need to use:*

- Find the total number of bird and mammal species in each ecosystem.

  49 + 94 = 143 (species in California)

  38 + 79 = 117 (species in Alaska)

- Compare the totals.

  143 is larger than 117, so California has the greater bird and mammal biodiversity.

4 *Check your answer:*

143 − 94 = 49 (mammal species in California)

117 − 79 = 38 (mammal species in Alaska)

Practice Problems

1. Compare the biodiversity of the coastal temperate rain forest ecosystems of Oregon with 55 mammal species and 99 permanent bird species to the British Columbia ecosystem with 68 mammal species and 116 permanent bird species.

2. Compare the biodiversity of lizards in hot North American deserts that have an average of 6.4 ground-dwelling lizards and 0.9 arboreal lizards with those in Africa that have an average of 9.8 ground-dwelling lizards and 3.5 arboreal lizards.

 Science Online

For more practice, visit booke.msscience.com/ math_practice

Figure 5 The passenger pigeon is an extinct species. These North American birds were over-hunted in the 1800s by frontier settlers who used them as a source of food and feathers.
Describe *other causes of extinction.*

What reduces biodiversity?

Flocks of thousands of passenger pigeons used to fly across the skies of North America. Few people today have ever seen one of these birds. The passenger pigeon, shown in **Figure 5,** has been extinct for almost 100 years. An **extinct species** is a species that was once present on Earth but has died out.

INTEGRATE Earth Science Extinction is a normal part of nature. The fossil record shows that many species have become extinct since life appeared on Earth. Extinctions can be caused by competition from other species or by changes in the environment. A mass extinction that occurred about 65 million years ago wiped out almost two-thirds of all species living on Earth, including the dinosaurs. This extinction, shown on the graph in **Figure 6,** occurred in the Mesozoic Era. It might have been caused by a huge meteorite that crashed into Earth's surface. Perhaps the impact filled the atmosphere with dust and ash that blocked sunlight from reaching Earth's surface. This event might have caused climate changes that many species could not survive. Mass extinctions eventually are followed by the appearance of new species that take advantage of the suddenly empty environment. After the dinosaurs disappeared, many new species of mammals appeared on Earth.

Reading Check *What are some causes of extinction?*

Loss of Species Not everyone agrees about the reason for the extinction of the dinosaurs. One thing is clear—human actions had nothing to do with it. Dinosaurs were extinct millions of years before humans were on Earth. Today is different. The rate of extinctions appears to be rising. From 1980 to 2000, close to 40 species of plants and animals in the United States became extinct. It is estimated that hundreds, if not thousands, of tropical species became extinct during the same 20-year period. Human activities probably contributed to most of these extinctions. As the human population grows, many more species could be lost.

Figure 6 This graph shows five mass extinctions in Earth's history. The mass extinctions appear as peaks on the graph.

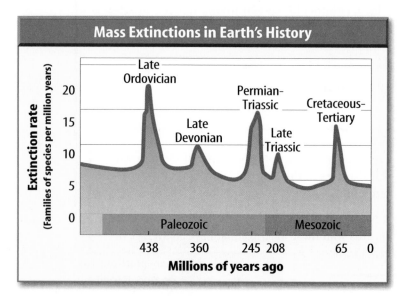

Mass Extinctions in Earth's History

Extinction rate (Families of species per million years)

Late Ordovician

Permian-Triassic

Cretaceous-Tertiary

Late Devonian

Late Triassic

Paleozoic   Mesozoic

438   360   245 208   65   0

Millions of years ago

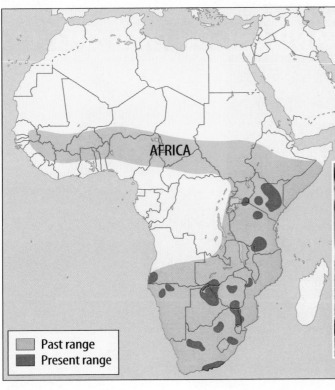

Past range
Present range

Figure 7 The African black rhinoceros is an endangered species, partly because people continue to kill these animals for their horns. As the map shows, this species once lived throughout southern, eastern, and central Africa.

Endangered Species To help prevent extinctions, it is important to identify species that could soon disappear. A species in danger of becoming extinct is classified as an **endangered species.** The African black rhinoceros, shown in **Figure 7,** is endangered. Rhinoceroses are plant eaters. They use their horns to battle each other for territory and to protect themselves against lions and other predators. For centuries, humans have considered rhinoceros horn to be a rare treasure. It is so valuable that poachers continue to hunt and kill these animals, even though selling rhinoceros horn is against international law. In 1970, there were about 100,000 black rhinoceroses in Africa. By the year 2000, fewer than 3,000 were left.

Threatened Species If a species is likely to become endangered in the near future, it is classified as a **threatened species.** The Australian koala, shown in **Figure 8,** is threatened. People once hunted koalas for their fur. In the 1930s, people realized the koala was in danger. Laws were passed that prohibited the killing of koalas, and the koala populations began to recover. Koalas rely on certain species of Australian eucalyptus trees for food and shelter. After the 1930s laws were passed, Australia's human population grew and the koala's habitat began to disappear. By the year 2000, nearly two-thirds of the koala's habitat had been lost to logging, agriculture, cities, and roads.

Figure 8 Koalas are related to kangaroos and opossums. The loss of their habitat threatens the future of wild koala populations.

Figure 9

The U.S. Fish and Wildlife Service lists more than 1,200 species of animals and plants as being at risk in the United States. Of these, about 20 percent are classified as threatened and 80 percent as endangered.

VERNAL POOL TADPOLE SHRIMP
This endangered species lives in the seasonal freshwater ponds of California's Central Valley. Pollution, urban sprawl, and other forces have destroyed 90 percent of the vernal pools in the valley. Loss of habitat makes the survival of these tiny creatures highly uncertain.

CALIFORNIA CONDOR
The endangered condor came close to extinction at the end of the twentieth century. Some condors have been raised in captivity and successfully released into the wild.

HALEAKALA SILVERSWORD One of the world's most spectacular plants, this threatened species is making a recovery. The plant shown here is blooming in Hawaii's Haleakala (ha lee ah kuh LAH) crater.

DESERT TORTOISE
The future of the threatened desert tortoise is uncertain. Human development is eroding tortoise habitat in the southwestern United States.

SOUTHERN SEA OTTER The threatened sea otter lives in shallow waters along the Pacific coast of the United States. For centuries, sea otters were hunted for their fur.

Habitat Loss In the Launch Lab at the beginning of this chapter, you probably observed that lawns and sidewalks have a lower biodiversity than sunlit woods or weed-covered lots. When people alter an ecosystem, perhaps by replacing a forest or meadow with pavement or a lawn, the habitats of some species may become smaller or disappear completely. If the habitats of many species are lost, biodiversity might be reduced.

Habitat loss is a major reason why species become threatened or endangered, as shown in **Figure 9,** or extinct. The Lake Erie water snake, shown in **Figure 10,** is classified as a threatened species because of habitat loss. These snakes live along the rocky shores of islands in Lake Erie. The islands are popular recreation spots. The development of boat docks and buildings in these areas has seriously reduced the amount of habitat available to the snakes. Also, these snakes are often killed by people who mistakenly think they are poisonous. Adult snakes range from 0.5 m to 1.0 m long, and they feed on fish, frogs, and salamanders. Because they have sharp teeth for capturing prey, they can bite. However, they are not dangerous to humans, especially if left undisturbed.

Conservation strategies for protecting the Lake Erie water snake include preserving its habitat by limiting development in some areas. Also, a public education program has been developed to inform people about this predator species and its importance in the Lake Erie ecosystem.

Mini LAB

Demonstrating Habitat Loss

Procedure
1. Put a small piece of **banana** in an **open jar.** Set the jar indoors near a place where food is prepared or fruit is thrown away.
2. Check the jar every few hours. When at least five fruit flies are in it, place a **piece of cloth or stocking** over the top of the jar and secure it with a **rubber band.**
3. Count and record the number of fruit flies in the jar every two days for three weeks.

Analysis
1. Explain why all the flies in the jar eventually die.
2. Use your results to hypothesize why habitat loss can reduce biodiversity.

Try at Home

Figure 10 The Lake Erie water snake has become a threatened species.

The building of boat docks, roads, and shopping centers is reducing the snake's habitat.

These snakes are not dangerous to humans, but many are killed by people who fear snakes.

Divided Habitats Biodiversity can be reduced when a habitat is divided by roads, cities, or farms. Small patches of habitat usually have less biodiversity than large areas. One reason for this is that large animals like mountain lions and grizzly bears require hunting territories that cover hundreds of square kilometers. If their habitat becomes divided, they are forced to move elsewhere.

Small habitat areas also make it difficult for species to recover from a disaster. Suppose a fire destroys part of a forest, and all the salamanders living there are destroyed. Later, after new trees have grown, salamanders from the undamaged part of the forest move in to replace those that were lost. But what if fire destroys a grove of trees surrounded by parking lots and buildings? Trees and salamanders perish. Trees might grow back but the salamanders might never return, because none live in the surrounding paved areas.

Figure 11 Introduced species can reduce or eliminate populations of native species in an ecosystem.

Introduced Species When species from another part of the world are introduced into an ecosystem, they can have a dramatic effect on biodiversity. An **introduced species** is a species that moves into an ecosystem as a result of human actions. Introduced species often have no competitors or predators in the new area, so their populations grow rapidly. Introduced species can crowd out or consume native species. **Native species** are the original organisms in an ecosystem.

In the early 1800s, European settlers brought goats with them to Santa Catalina Island off the coast of California. The goats overgrazed the land, completely eliminating 48 of the native plant species and exposing patches of bare soil. The bare soil provided a place for hardy, introduced weeds to take root. The weeds crowded out native species. Late in the twentieth century, after goats were removed from the island, some native plant species began to recover. **Figure 11** shows introduced species that have reduced biodiversity in other ecosystems.

The ruffe was introduced into the Great Lakes by ships from other parts of the world. It quickly takes over habitat and food sources used by native fish.

Purple loosestrife was brought to North America from Europe and Asia in the 1800s. Dense patches of loosestrife grow and crowd out the native plants that some animals need for food and shelter.

Pollution

Biodiversity also is affected by pollution of land, water, or air. Soil that is contaminated with oil, chemicals, or other pollutants can harm plants or limit plant growth. Because plants provide valuable habitat for many species, any reduction in plant growth can limit biodiversity.

Figure 12 Acid rain can damage the leaves and other tissues of trees. **Explain** *how this damage can affect other organisms.*

Water Pollution Water-dwelling organisms are easily harmed by pesticides, chemicals, oil, and other pollutants that contaminate the water. Water pollutants often come from factories, ships, or runoff from roads, lawns, and farms. Waterways also can be polluted when people dispose of wastes improperly. For example, excess water from streets and roads runs into storm drains during rainstorms. This water usually flows untreated into nearby waterways. Storm drains should never be used to dispose of used motor oil, paints, solvents, or other liquid wastes. These pollutants can kill aquatic plants, fish, frogs, insects, and the organisms they depend on for food.

Air Pollution A form of water pollution known as acid rain is caused by air pollution. **Acid rain** forms when sulfur dioxide and nitrogen oxide released by industries and automobiles combine with water vapor in the air. As **Figure 12** shows, acid rain can have serious effects on trees. It washes calcium and other nutrients from the soil, making the soil less fertile. One tree species that is particularly vulnerable to acid rain is the sugar maple. Many sugar maple trees in New England and New York have suffered major damage from acid rain. Acid rain also harms fish and other organisms that live in lakes and streams. Some lakes in Canada have become so acidic that they have lost almost all of their fish species. In the United States, 14 eastern states have acid rain levels high enough to harm aquatic life.

Air pollution from factories, power plants, and automobiles can harm sensitive tissues of many organisms. For example, air pollution can damage the leaves or needles of some trees. This can weaken them and make these trees less able to survive diseases, attacks by insects and other pests, or environmental stresses such as drought or flooding.

Mini LAB

Modeling the Effects of Acid Rain

Procedure
1. Soak **50 mustard seeds** in **water** and **another 50 mustard seeds** in **vinegar** (an acid) for 24 hours.
2. Wrap each group of seeds in a **moist paper towel.**
3. Put them into separate, **self-sealing plastic bags.** Seal and label each bag.
4. After 3 days, open each bag, examine the seeds, and record the number that show evidence of growth.

Analysis
1. Describe the effect of an acid on mustard seed growth.
2. Explain how acid rain could affect plant biodiversity.

Global Warming Carbon dioxide gas (CO_2) is released into the atmosphere when wood, coal, gas, or any other fuel is burned. People burn large amounts of fuel, and this is contributing to an increase in the percentage of CO_2 in the atmosphere. An increase in CO_2 could raise Earth's average temperature by a few degrees. This average temperature rise, called global warming, might lead to climatic changes that could affect biodiversity. For example, portions of the polar ice caps could melt, causing floods in coastal ecosystems around the world.

Ozone Depletion The atmosphere includes the ozone layer—ozone gas that is about 15 km to 30 km above Earth's surface. It protects life on Earth by preventing damaging amounts of the Sun's ultraviolet (UV) radiation from reaching Earth's surface. Scientists have discovered that the ozone layer is gradually becoming thinner. The thinning of the ozone layer is called **ozone depletion.** This depletion allows increased amounts of UV radiation that can harm living organisms to reach Earth's surface. For humans, it could mean more cases of skin cancer. Ozone depletion occurs over much of Earth. Data collected in the late 1990s indicate that ozone levels over the United States, for example, had decreased by five to ten percent since the 1970s.

section 1 review

Summary

The Variety of Life

- Biodiversity is a measurement of the number of species living in an area.
- In warmer climates, biodiversity usually is greater than in cooler climates.

Why is biodiversity important?

- Biodiversity can improve crops and provide materials for furniture, clothing, and medicine.

What reduces biodiversity?

- Extinction is a normal part of nature that reduces biodiversity.
- Climate changes, habitat loss, and introduced species can decrease biodiversity.

Pollution

- Ozone depletion can harm organisms by increasing their exposure to UV radiation.

Self Check

1. **Describe** how endangered species are different from extinct species.
2. **Explain** why habitat loss and habitat division are threats to biodiversity.
3. **Explain** how introduced species threaten biodiversity.
4. **Think Critically** Aluminum, commonly found in soil, is toxic to fish. It damages the ability of gills to absorb oxygen. Acid rain can wash aluminum out of the soil and into nearby waterways. Explain how acid rain could affect biodiversity in a pond.

Applying Skills

5. **Identify and Manipulate Variables, Constants, and Controls** Lack of calcium in the soil can damage trees. Design an experiment to test the hypothesis that acid rain is removing calcium from forest soil.

Oily Birds

You probably have seen news clips of waterbirds covered with oil after an oil spill. Nearly four billion kilograms of oil are spilled from ships or dumped into the ocean each year. Oil floats on water, and can coat the feathers of waterbirds. These birds depend on their feathers for insulation from the cold. Oiled feathers lose their insulating qualities, so oiled birds are exposed to the cold. Oil-coated feathers also prevent the birds from flying and can even cause them to drown. Rescue teams try to clean off the oil, but as you'll find out in this lab, this is a difficult task.

◉ Real-World Question

How difficult is it to clean oil from bird feathers?

Goals
■ **Experiment** with different methods for cleaning oil from bird feathers.

Materials
beakers (2)

bird feathers
   (white or light colored)

vegetable oil or olive oil

red, blue, and
   green food coloring

paper towels

water

cotton balls

cotton swabs

sponge

toothbrush

dish soap

Safety Precautions

◉ Procedure

1. Fill a beaker with vegetable oil and add several drops of red, blue, and green food coloring. Submerge your feathers in the oil for several minutes.

2. Lay paper towels on a table. Remove the feathers from the oil and allow the excess oil to drip back into the beaker. Lay the feathers on the paper towels.

3. Using another paper towel, blot the oil from one feather. Run your fingers over the feather to determine if the oil is gone.

4. Using the cotton swabs and cotton balls, wipe the oil from a second feather. Check the feather for oil when finished.

5. Using a toothbrush, brush the oil from a third feather. Check it for oil when you are finished.

6. Use a sponge to clean the fourth feather. Check it for oil when you are finished.

7. Fill the second beaker with water and add several drops of dish soap. Soak the fifth oil-soaked feather in the solution. Check the feather for oil after it has soaked for several minutes.

◉ Conclude and Apply

1. **Identify** the method or methods that best removed the oil from the feathers.

2. **Describe** the condition of the feathers after you finished cleaning them. How would the condition of these feathers affect birds?

3. **Infer** why rescuers would be hesitant to use soap when cleaning live birds.

4. **Infer** how oil pollution affects water mammals such as otters and seals.

𝒞ommunicating Your Data

Share your results with your classmates. Decide which methods you would use to try to clean a live bird.

Conservation Biology

as you read

What You'll Learn

- **Identify** several goals of conservation biology.
- **Recommend** strategies to prevent the extinction of species.
- **Explain** how an endangered species can be reintroduced into its original habitat.

Why It's Important

Conservation biology provides ways to preserve threatened and endangered species.

🔁 Review Vocabulary

habitat: the place or environment where a plant or animal naturally or normally lives and grows

New Vocabulary

- conservation biology
- habitat restoration
- captive population
- reintroduction program

Protecting Biodiversity

The study of methods for protecting biodiversity is called **conservation biology.** Conservation biologists develop strategies to prevent the continuing loss of members of a species. Conservation strategies must be based on a thorough understanding of the principles of ecology. Because human activities are often the reason why a species is at risk, conservation plans also must take into account the needs of the human population. The needs of humans and of other species often conflict. It can be difficult to develop conservation plans that satisfy both. Conservation biologists must consider law, politics, society, and economics, as well as ecology, when they look for ways to conserve Earth's biodiversity.

The Florida manatee, shown in **Figure 13,** is an endangered species. Manatees are plant-eating mammals that live in shallow water along the coasts of Florida and the Carolinas. These animals swim slowly, occasionally rising to the surface to breathe. They swim below the surface, but not deep enough to avoid the propellers of powerboats. Many manatees have been injured or killed by powerboat propellers. Boaters cannot always tell when manatees are nearby, and it is difficult to enforce speed limits to protect the manatees. Also, the manatee's habitat is being affected by Florida's growing human population. More boat docks are being built, which can add to the number of boats in the water. Water pollution from boats, roads, and coastal cities is also a problem.

✓ Reading Check *How do humans affect manatee habitats?*

Figure 13 Conservation biologists are working to protect the endangered Florida manatee by limiting habitat loss, reducing pollution, and encouraging boaters to obey speed limits. Rescued manatees are being rehabilitated at the Columbus Zoo in Ohio.

Figure 14 The American bald eagle and American alligator once were listed as endangered. **Infer** *why the American bald eagle and American alligator were taken off the endangered species list.*

Conservation Biology at Work

Almost every conservation plan has two goals. The first goal is to protect a species from harm. The second goal is to protect the species' habitat. Several conservation strategies can be used to meet these goals.

Legal Protections Laws can be passed to help protect a species and its habitat. In the early 1970s, people became concerned about the growing number of species extinctions occurring in the United States. In response to this concern, Congress passed the U.S. Endangered Species Act of 1973. This act makes it illegal to harm, collect, harass, or disturb the habitat of any species on the endangered or threatened species lists. The act also prevents the U.S. government from spending money on projects that would harm these species or their habitats. The Endangered Species Act has helped several species, including those shown in **Figure 14,** recover from near extinction.

The United States and other countries worldwide have agreed to work together to protect endangered or threatened species. In 1975, The Convention on International Trade in Endangered Species of Wild Fauna and Flora, known as CITES, came into existence. One of its goals is to protect certain species by controlling or preventing international trade in these species or any part of them, such as elephant ivory. About 5,000 animal species and 25,000 plant species are protected by this agreement.

Topic: Endangered Species
Visit booke.msscience.com for Web links to information about endangered species.

Activity Research an endangered or threatened species. Write a description of the species and explain why it is considered in danger.

INTEGRATE
Social Studies

Marine Biodiversity The marine fish catch has an estimated value of $80 billion per year. Marine environments also provide us with raw materials and medicines. Other uses of marine environments include research, eco-tourism, and recreation. Research and then describe in your Science Journal what is being done to protect marine environments.

Habitat Preservation Even if a species is protected by law, it cannot survive unless its habitat also is protected. Conservation biology often focuses on protecting habitats, or even whole ecosystems. One way to do this is to create nature preserves, such as national parks and protected wildlife areas.

The United States established its first national park—Yellowstone National Park—in 1872. At that time, large animals like grizzly bears, elk, and moose ranged over much of North America. These animals roam across large areas of land in search of food. If their habitat becomes too small, they cannot survive. The grizzly bear, for example, requires large quantities of food each day. To feed itself, a grizzly needs a territory of several hundred square kilometers. Without national parks and wildlife areas, some animals would be far fewer in number than they are today.

Wildlife Corridors The successful conservation of a species like the grizzly bear requires enormous amounts of land. However, it is not always possible to create large nature preserves. One alternative is to link smaller parks together with wildlife corridors. Wildlife corridors allow animals to move from one preserve to another without having to cross roads, farms, or other areas inhabited by humans. As **Figure 15** shows, wildlife corridors are part of the strategy for saving the endangered Florida panther. A male panther needs a territory of 712 km^2 or more, which is larger than many of the protected panther habitats.

Figure 15 Some wildlife corridors (left) allow animals to pass safely under roads and highways. This map shows the types of habitat where most populations of Florida panthers are found. Notice the major highways that cross these areas.

Panthers
Florida Habitats
Pinelands
Freshwater marsh
Cypress swamp
Prairies
Hardwood swamp
Swamp & Coastal

Habitat Restoration Habitats that have been changed or harmed by human activities often can be restored. **Habitat restoration** is the process of taking action to bring a damaged habitat back to a healthy condition.

Rhode Island's Narragansett Bay was once an important fishing area. It supported populations of flounder, bay scallops, blue crabs, and other important food species. Eelgrass, which is an underwater plant that grows in shallow parts of the bay, provides habitat for the young of many of these species. As the human population of Rhode Island grew, most of the bay's eelgrass beds disappeared. Fish and shellfish populations declined. People who lived in the area knew that something needed to be done. Students in Rhode Island schools have helped restore eelgrass habitat by growing thousands of eelgrass seedlings for transplanting into the bay, as shown in **Figure 16.**

Figure 16 By restoring eelgrass habitat, conservation biologists hope to preserve populations of shellfish and fish in Narragansett Bay. **Describe** *how the restoration of eelgrass habitat will help preserve populations of shellfish and fish.*

✓ Reading Check *When is habitat restoration used?*

Wildlife Management Preserving or restoring a habitat does not mean that all the species living there are automatically protected from harm. Park rangers, guards, and volunteers often are needed to manage the area. In South Africa, guards patrol wildlife parks to prevent poachers from killing elephants for their tusks. Officials protect the mountain gorillas of Rwanda by limiting the number of visitors allowed to see them. Some wildlife preserves allow no visitors other than biologists who are studying the area.

Hunters and wildlife managers often work together to maintain healthy ecosystems in parks and preserves. People usually are not allowed to hunt or fish in a park unless they purchase a hunting or fishing license. The sale of licenses provides funds for maintaining the wildlife area. It also helps protect populations from overhunting. For example, licenses may limit the number of animals a hunter is allowed to take or may permit hunting only during seasons when a species is not reproducing. Hunting regulations also can help prevent a population from becoming too large for the area.

Captive Populations The Arabian oryx, shown in **Figure 17,** is native to desert lands of the Arabian Peninsula. People hunt the oryx for its horns and hide. Hunting increased during the twentieth century, when four-wheel-drive vehicles became available to hunters. In 1962, to prevent the species from becoming extinct, several oryx were taken to the Phoenix Zoo in Arizona. By 1972, no more wild oryx lived in the desert. A **captive population** is a population of organisms that is cared for by humans. The captive population of oryx in Arizona did well. Their numbers increased, and hundreds of Arabian oryx now live in zoos all over the world.

Figure 17 The Arabian oryx was saved from extinction because a captive population was created before the wild population disappeared.

Keeping endangered or threatened animals in captivity can help preserve biodiversity. It is not ideal, however. It can be expensive to provide proper food, adequate space, and the right kind of care. Also, captive animals sometimes lose their wild behaviors. If that happens, they might not survive if they're returned to their native habitats. The best approach is to preserve the natural habitats of these organisms so they can survive on their own.

Reintroduction Programs In some cases, members of captive populations can be released into the wild to help restore biodiversity. In the 1980s, wildlife managers began reintroducing captive oryx into their desert habitat. **Reintroduction programs** return captive organisms to an area where the species once lived. Programs like this can succeed only if the factors that caused the species to become endangered are removed. In this case, the reintroduced oryx must be protected from illegal hunting.

Plants also can be reintroduced into their original habitats. The island of St. Helena in the South Atlantic was home to the rare St. Helena ebony tree. In the 1500s, European settlers introduced goats to the island. Settlers also cut trees for firewood and timber. Goats overgrazed the land, and native trees were replaced by introduced species. By 1850, the St. Helena ebony was thought to be extinct. In 1980, two small ebony trees were found on the island, high on a cliff. Small branches were cut from these trees and rooted in soil. These cuttings have been used to grow thousands of new ebony trees. The new trees are being replanted all over the island.

Science Online

Topic: Reintroduction Programs

Visit booke.msscience.com for Web links to information about reintroduction programs.

Activity Find out about a reintroduction program and write a summary of the program's progress.

Seed Banks Throughout the world, seed banks have been created to store the seeds of many endangered plant species. If any of these species become extinct in the wild, the stored seeds can be used to reintroduce them to their original habitats.

Relocation Reintroduction programs do not always involve captive populations. In fact, reintroductions are most successful when wild organisms are transported to a new area of suitable habitat. The brown pelican, shown in **Figure 18,** was once common along the shores of the Gulf of Mexico. Pelicans eat fish that eat aquatic plants. In the mid-twentieth century, DDT—a pesticide banned in 1972 in the U.S.—was used widely to control insect pests. It eventually ended up in the food that pelicans ate. Because of DDT, the pelican's eggshells became so thin that they would break before the chick inside was ready to hatch. Brown pelicans completely disappeared from Louisiana and most of Texas. In 1971, 50 of these birds were taken from Florida to Louisiana. Since then, the population has grown. In the year 2000, more than 7,000 brown pelicans lived in Louisiana and Texas.

Figure 18 Brown pelicans from Florida were relocated successfully to Louisiana.

Reading Check *What two things led to the recovery of the brown pelican?*

section 2 review

Summary

Protecting Biodiversity

- Conservation biology deals with protecting biodiversity.
- Conservation plans must take into account the needs of the human population because human activities often are the reason species are at risk.

Conservation Biology at Work

- Conservation plans are designed to protect a species from harm and protect their habitat.
- Laws, habitat preservation, wildlife corridors, and habitat restoration are techniques used to protect a species and its habitat.
- Sometimes captive or wild populations can be released into a new area to help restore biodiversity.

Self Check

1. **Explain** how the U.S. Endangered Species Act helps protect species.
2. **Describe** what factors can make it difficult to reintroduce captive animals into their natural habitats.
3. **Think Critically** As a conservation biologist, you are helping to restore eelgrass beds in Narragansett Bay. What information will be needed to prepare a conservation plan?

Applying Skills

4. **Recognize Cause and Effect** Imagine creating a new wildlife preserve in your state. Identify uses of the area that would allow people to enjoy it without harming the ecosystem. What kinds of uses might damage the habitat of species that live there?

Biodiversity and the Health of a Plant Community

Real-World Question

Some plant diseases are carried from plant to plant by specific insects and can spread quickly throughout a garden. If a garden only has plants that are susceptible to one of these diseases, the entire garden can be killed when an infection occurs. One such disease—necrotic leaf spot—can infect impatiens and other garden plants. Could biodiversity help prevent the spread of necrotic leaf spot? A simulation can help you answer this question. How does biodiversity affect the spread of a plant disease?

Goals

- **Record data** on the spread of a plant disease.
- **Compare** the spread of disease in communities with different levels of biodiversity.
- **Predict** a way to increase a crop harvest.

Materials

plain white paper
colored paper (red, orange, yellow, green, blue, black)
ruler
scissors
pens
die

Safety Precautions

Use care when cutting with scissors.

Procedure

1. On a piece of white paper, draw a square that measures 10 cm on each side. This represents a field. In the square, make a grid with five equal rows and five equal columns, as shown. Number the outer cells from 1 through 16, as shown. The inner cells are not numbered.

2. Cut 1.5-cm × 1.5-cm tiles from colored paper. Cut out 25 black tiles, 25 red tiles, 10 orange tiles, 10 yellow tiles, 5 green tiles, and 5 blue tiles. The black tiles represent a plant that has died from a disease. The other colors represent different plant species. For example, all the red tiles represent one plant species, all the blue tiles are another species, and each remaining color is a different species.

3. There are four rounds in this simulation. For each round, randomly distribute one plant per square as instructed. Roll the die once for each square, proceeding from square 1 to square 2, and so on through square 16.

Round 1—High Biodiversity
Distribute five red, five orange, five yellow, five green, and five blue plants in the field.

Round 2—Moderate Biodiversity
Distribute ten orange, ten yellow, and five green plants in the field.

Round 3—Low Biodiversity
Distribute all 25 of the red tiles in the field.

1	2	3	4	5
16				6
15				7
14				8
13	12	11	10	9

Round 4—Challenge

You want a harvest of as many red plants as possible. Decide how many red plants you will start with and whether you will plant any other species. Strategically place your tiles on the board. Follow steps 3 through 5 of the procedure.

4. To begin, place a black tile over half of square 1. Roll the die and use the Disease Key to see if the plant in square 1 is infected. If it is, cover it with the black tile. If not, remove the black tile. For example, suppose square 1 has an orange plant on it. Place a black tile over half of the square and roll the die. If you roll a 1, 3, 4, or 5, the plant does not get the disease, so remove the black tile. Proceed through all the squares until you have rolled the die 16 times.

5. If two plants of the same species are next to each other, the disease will spread from one to the other. Suppose squares 2 and 3 contain red plants. If you roll a 1 for square 2, the red plants on squares 2 and 3 die. In this case, the die does not need to be rolled for square 3. Similarly, if the inner square next to square 2 also contains a red plant, it dies and should be covered with a black tile.

6. For each round, record in a data table the number of tiles of each species you started with and the number of each species left alive at the end of the round.

Disease Key	
Roll of Die	Infected Color
1	Red
2	Orange
3	Yellow
4	Blue
5	Green
6	All colors

Round _____		
Color of Species	Number at Start	Number at End
Red		
Orange	Do not write in	
Yellow	this book.	
Blue		
Green		
TOTAL		

Analyze Your Data

Identify which round had the lowest number of survivors.

Conclude and Apply

1. **Infer** why a high biodiversity helps a community survive.
2. **Explain** why organic farmers often grow crops of different species in the same field.

Communicating Your Data

Share your results from Round 4 with your classmates. Decide which arrangement and number of red tiles yields the best harvest.

RAIN FOREST Troubles

Large areas of Brazil's rain forests are being burned to make way for homes, farms, and ranches.

Tropical rain forests are located on both sides of the equator. The average temperature in a rain forest is about 25°C and doesn't vary much between day and night. Rain forests receive, on average, between 200 cm and 400 cm of rain each year. Tropical rain forests contain the highest diversity of life on Earth. Scientists estimate that there are about 30 million plant and animal species on this planet—and at least half of them live in rain forests!

Tropical rain forests impact all of our lives every day. Plants in the forest remove carbon dioxide from the atmosphere and give off oxygen. Some of these plants are already used in medicines, and many more are being studied to determine their usefulness.

Humans destroy as many as 20 million hectares of tropical rain forests each year. Farmers who live in tropical areas cut the trees to sell the wood and farm the land. After a few years, the crops use up the nutrients in the soil and more land must be cleared for farming. The logging and mining industries also contribute to the loss of valuable rain forest resources.

Conservationists suggest a number of ways to protect the rain forests.

- Have governments protect rain forest land by establishing parks and conservation areas.
- Reduce the demand for industrial timber and paper.
- Require industries to use different tree-removal methods that cause less damage.
- Teach farmers alternative farming methods to decrease the damage to rain forest land.
- Offer money or lower taxes to farmers and logging companies for not cutting down trees on rain forest land.

Carrying out these plans will be difficult. The problem extends far beyond the local farmers—the whole world strips rain forest resources in its demand for products such as food ingredients and lumber for building homes.

This tropical rain forest in Costa Rica is untouched by human development.

Present Prepare a multimedia presentation for an elementary class. Teach the class what a rain forest is, where rain forests are located, and how people benefit from rain forests.

Science online

For more information, visit booke.msscience.com/time

Reviewing Main Ideas

Section 1 Biodiversity

1. A measure of biodiversity is the number of species present in an ecosystem.

2. In general, biodiversity is greater in warm, moist climates than in cold, dry climates.

3. Extinction occurs when the last member of a species dies.

4. Habitat loss, pollution, overhunting, and introduced species can cause a species to become threatened or endangered.

5. Global warming and ozone depletion could affect biodiversity.

Section 2 Conservation Biology

1. Conservation biology is the study of methods for protecting Earth's biodiversity.

2. The Endangered Species Act of 1973 preserves biodiversity by making it illegal to harm threatened or endangered species.

3. Habitat preservation, habitat restoration, and wildlife management strategies can be used to preserve species.

4. Reintroduction programs can be used to restore a species to an area where it once lived.

Visualizing Main Ideas

Copy and complete the following concept map using the following terms:
habitat restoration, captive populations, relocation, *and* endangered species.

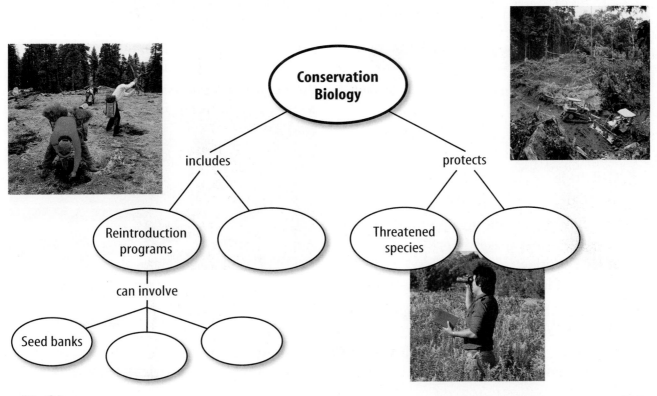

Using Vocabulary

acid rain p. 135
biodiversity p. 126
captive population p. 142
conservation
    biology p. 138
endangered species p. 131
extinct species p. 130

habitat restoration p. 141
introduced species p. 134
native species p. 134
ozone depletion p. 136
reintroduction
    program p. 142
threatened species p. 131

Explain the differences in the vocabulary words given below. Then explain how the words are related.

1. habitat restoration—threatened species
2. endangered species—extinct species
3. biodiversity—conservation biology
4. biodiversity—captive population
5. acid rain—ozone depletion
6. introduced species—endangered species
7. native species—introduced species
8. biodiversity—extinct species
9. reintroduction program—native species

Checking Concepts

Choose the word or phrase that best answers the question.

10. Which state probably has the greatest biodiversity?
    A) Florida   **C)** New Hampshire
    B) Maine   **D)** West Virginia

11. Which of these is the greatest threat to the Lake Erie water snake?
    A) air pollution   **C)** ozone
    B) habitat loss   **D)** introduced species

12. Which of the following gases might contribute to global warming?
    A) oxygen   **C)** carbon dioxide
    B) nitrogen   **D)** neon

13. Which of the following can reduce biodiversity?
    A) divided habitats
    B) habitat loss
    C) introduced species
    D) all of these

14. Jen has a tank with 20 guppies. June has a tank with 5 guppies, 5 mollies, 5 swordtails, and 5 platys. Which of the following is true?
    A) June has more fish than Jen.
    B) June's tank has more biodiversity than Jen's tank.
    C) Jen has more fish than June.
    D) Jen's tank has more biodiversity than June's tank.

15. What would you call a species that had never lived on an island until it was brought there by people?
    A) introduced   **C)** native
    B) endangered   **D)** threatened

16. Which of the following conservation strategies involves passing laws to protect species?
    A) habitat restoration
    B) legal protections
    C) reintroduction programs
    D) captive populations

Use the photo below to answer question 17.

17. The animals above are examples of a(n)
    A) extinct species.
    B) endangered species.
    C) lost species.
    D) introduced species.

Science Online   booke.msscience.com/vocabulary_puzzlemaker

Thinking Critically

18. Infer What action(s) could people take to help reduce the types of air pollution that might contribute to global warming?

19. Describe what conservation strategies would most likely help preserve a species whose members are found only in zoos.

Use the following diagrams to answer question 20.

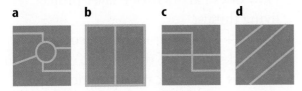

20. Draw a Conclusion Four designs for a road system in a national park are shown above. Which arrangement best would avoid dividing this habitat into pieces? Explain.

21. Infer Why might the protection of an endangered species create conflict between local residents and conservation biologists?

22. Explain why habitat loss is the most serious threat to biodiversity.

23. Concept Map Draw an events-chain concept map using the following terms: *species*, *extinct species*, *endangered species*, and *threatened species*.

24. Form a Hypothesis In divided wildlife areas, large animals have a greater chance of becoming extinct than other organisms do. Suggest a hypothesis to explain why this is true.

25. Compare and Contrast Explain how legal protections and wildlife management are similar and different.

26. Recognize Cause and Effect Design a wildlife management plan to allow deer hunting in a state park without damaging the ecosystem.

Performance Activities

27. Poster Use images from magazines to create a display about how humans threaten Earth's biodiversity.

Applying Math

Use the following table to answer questions 28–30.

Wildlife Habitat Lost		
Country	Area Lost (km^2)	Area Remaining (km^2)
Ethiopia	770,700	330,300
Vietnam	265,680	66,420
Indonesia	708,751	746,860

28. Habitat Loss Use the data in the table to make a bar graph.

29. Original Habitat Using the table above, calculate the beginning amount of habitat in each country.

30. Remaining Habitat Use the table above to calculate the proportion of wildlife habitat area remaining in Ethiopia, Vietnam, and Indonesia.

31. Extinction Since 1600, the number of known extinctions includes 29 fish, 2 amphibians, 23 reptiles, 116 birds, and 59 mammals. How many vertebrates are known to have gone extinct since 1600?

32. Biodiversity A pond contains 6 species of fish, 3 species of amphibians, and 2 species of reptiles. A new species of fish is released into the pond resulting in the loss of 2 fish species and 2 amphibian species. How many species are in the pond before the new species of fish is released and how many species are in the pond after it is released?

33. Introduced Species After the brown tree snake was introduced to the island of Guam, 9 of 11 native forest-dwelling bird species became extinct. What proportion of native forest-dwelling bird species remain?

Part 1 | Multiple Choice

Record your answers on the answer sheet provided by your teacher or on a sheet of paper.

Use the illustration below to answer questions 1 and 2.

Plan A

Plan B

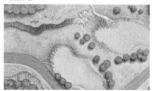

1. A road is being built through an area near two lakes and a construction team must chose between Plan A and Plan B. A conservation biologist evaluates how each plan would affect biodiversity in the ecosystem that includes the lakes and urges the team to follow Plan A. Why?
   A. It introduces new species.
   B. It does not divide the aquatic habitat into small areas.
   C. It makes the road easier to drive along and view animals.
   D. It helps keep more water in both of the lakes.

2. Which species is least likely to be affected if the construction team follows Plan B?
   A. water snakes   **C.** algae
   B. frogs   **D.** birds

3. Where would you expect to find more biodiversity?
   A. wheat field
   B. polluted stream
   C. lawn
   D. coral reef

Test-Taking Tip

Test Day Preparation Avoid rushing on test day. Prepare your clothes and test supplies the night before. Wake up early and arrive at school on time on test day.

4. What is used to link small parks together so animals can move easily from one to another?
   A. habitat restoration
   B. relocation
   C. wildlife corridors
   D. seed banks

5. How does Earth's ozone layer protect living things?
   A. keeps Earth's temperature constant
   B. blocks much of the Sun's UV radiation
   C. stops polar ice caps from melting
   D. prevents acid rain from falling to Earth

Use the illustration below to answer questions 6 and 7.

6. The small mollusks shown above are zebra mussels, a native species in the Caspian Sea in Asia. They were accidentally brought to the Great Lakes in the ballast water of boats. How would zebra mussels be described in North America?
   A. reintroduced species
   B. introduced species
   C. endangered species
   D. native species

7. Zebra mussels grow more rapidly than mussels native to the Great Lakes and cause the native mussels to die out. Which of the following best describes the problem caused by zebra mussels?
   A. They are decreasing biodiversity.
   B. They are causing divided habitats.
   C. They are polluting the water.
   D. They are causing global warming.

Record your answers on the answer sheet provided by your teacher or on a sheet of paper.

8. What law makes it illegal to harm, collect, harass, or disturb the habitat of the American bald eagle and the American alligator?

9. What is the study of the methods for protecting biodiversity called?

10. What are two goals of most conservation plans?

11. What are two gases released from automobiles and industries that can combine with water vapor in the air to cause acid rain?

Use the illustration below to answer questions 12 and 13.

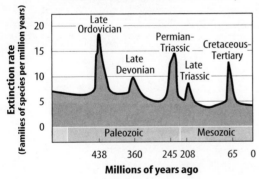

Mass Extinctions in Earth's History

12. The graph shows mass extinctions that have occurred in Earth's history. How many mass extinctions have occurred?

13. Did human activities contribute to any of these extinctions? Explain your answer.

14. Why is biodiversity important to the field of medicine?

15. Brown pelicans once completely disappeared from Louisiana and most of Texas. Today, thousands of them live in these areas. What type of conservation program caused this change?

Record your answers on a sheet of paper.

16. How can hunters and wildlife managers work together to maintain ecosystems in parks and preserves?

17. Global warming and ozone depletion both involve changes to Earth's atmosphere. Compare and contrast the problems these two processes could cause to the environment.

18. Explain why just protecting a species like the grizzly bear with a law may not be enough to conserve this species.

19. Describe what people do that causes the release of large amounts of carbon dioxide (CO_2) into the atmosphere. What problems might an increase in CO_2 cause for people?

Use the photo below to answer questions 20 and 21.

20. Explain why the small round sign placed beside this storm drain says "No Dumping."

21. Describe other ways that pollutants can enter water. How is biodiversity affected by pollutants?

22. Why are captive populations not an ideal way to protect endangered species?

23. Compare and contrast legal protection, habitat restoration, and seed banks as programs used to prevent the loss of species from an ecosystem.

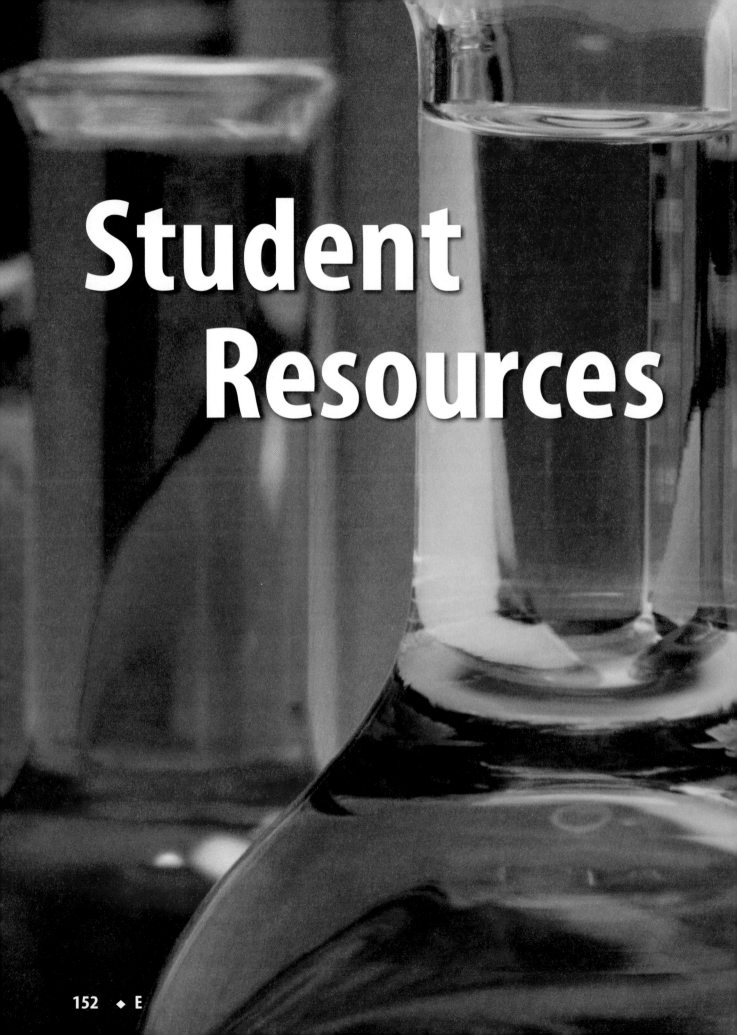

Student
Resources

CONTENTS

Scientific Methods

Scientists use an orderly approach called the scientific method to solve problems. This includes organizing and recording data so others can understand them. Scientists use many variations in this method when they solve problems.

Identify a Question

The first step in a scientific investigation or experiment is to identify a question to be answered or a problem to be solved. For example, you might ask which gasoline is the most efficient.

Gather and Organize Information

After you have identified your question, begin gathering and organizing information. There are many ways to gather information, such as researching in a library, interviewing those knowledgeable about the subject, testing and working in the laboratory and field. Fieldwork is investigations and observations done outside of a laboratory.

Researching Information Before moving in a new direction, it is important to gather the information that already is known about the subject. Start by asking yourself questions to determine exactly what you need to know. Then you will look for the information in various reference sources, like the student is doing in **Figure 1.** Some sources may include textbooks, encyclopedias, government documents, professional journals, science magazines, and the Internet. Always list the sources of your information.

Figure 1 The Internet can be a valuable research tool.

Evaluate Sources of Information Not all sources of information are reliable. You should evaluate all of your sources of information, and use only those you know to be dependable. For example, if you are researching ways to make homes more energy efficient, a site written by the U.S. Department of Energy would be more reliable than a site written by a company that is trying to sell a new type of weatherproofing material. Also, remember that research always is changing. Consult the most current resources available to you. For example, a 1985 resource about saving energy would not reflect the most recent findings.

Sometimes scientists use data that they did not collect themselves, or conclusions drawn by other researchers. This data must be evaluated carefully. Ask questions about how the data were obtained, if the investigation was carried out properly, and if it has been duplicated exactly with the same results. Would you reach the same conclusion from the data? Only when you have confidence in the data can you believe it is true and feel comfortable using it.

Interpret Scientific Illustrations As you research a topic in science, you will see drawings, diagrams, and photographs to help you understand what you read. Some illustrations are included to help you understand an idea that you can't see easily by yourself, like the tiny particles in an atom in **Figure 2.** A drawing helps many people to remember details more easily and provides examples that clarify difficult concepts or give additional information about the topic you are studying. Most illustrations have labels or a caption to identify or to provide more information.

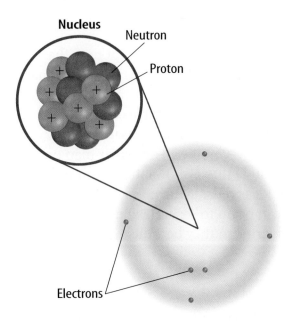

Figure 2 This drawing shows an atom of carbon with its six protons, six neutrons, and six electrons.

Concept Maps One way to organize data is to draw a diagram that shows relationships among ideas (or concepts). A concept map can help make the meanings of ideas and terms more clear, and help you understand and remember what you are studying. Concept maps are useful for breaking large concepts down into smaller parts, making learning easier.

Network Tree A type of concept map that not only shows a relationship, but how the concepts are related is a network tree, shown in **Figure 3.** In a network tree, the words are written in the ovals, while the description of the type of relationship is written across the connecting lines.

When constructing a network tree, write down the topic and all major topics on separate pieces of paper or notecards. Then arrange them in order from general to specific. Branch the related concepts from the major concept and describe the relationship on the connecting line. Continue to more specific concepts until finished.

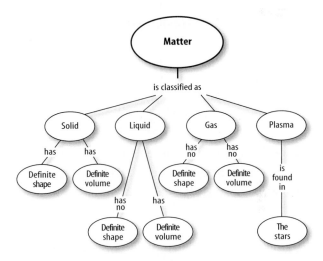

Figure 3 A network tree shows how concepts or objects are related.

Events Chain Another type of concept map is an events chain. Sometimes called a flow chart, it models the order or sequence of items. An events chain can be used to describe a sequence of events, the steps in a procedure, or the stages of a process.

When making an events chain, first find the one event that starts the chain. This event is called the initiating event. Then, find the next event and continue until the outcome is reached, as shown in **Figure 4.**

Initiating Event

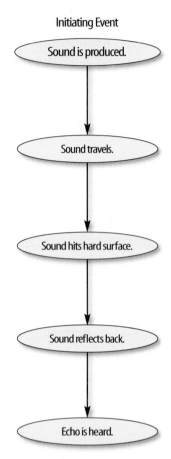

Figure 4 Events-chain concept maps show the order of steps in a process or event. This concept map shows how a sound makes an echo.

Cycle Map A specific type of events chain is a cycle map. It is used when the series of events do not produce a final outcome, but instead relate back to the beginning event, such as in **Figure 5.** Therefore, the cycle repeats itself.

To make a cycle map, first decide what event is the beginning event. This is also called the initiating event. Then list the next events in the order that they occur, with the last event relating back to the initiating event. Words can be written between the events that describe what happens from one event to the next. The number of events in a cycle map can vary, but usually contain three or more events.

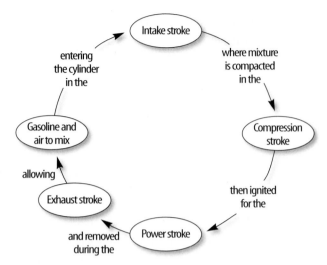

Figure 5 A cycle map shows events that occur in a cycle.

Spider Map A type of concept map that you can use for brainstorming is the spider map. When you have a central idea, you might find that you have a jumble of ideas that relate to it but are not necessarily clearly related to each other. The spider map on sound in **Figure 6** shows that if you write these ideas outside the main concept, then you can begin to separate and group unrelated terms so they become more useful.

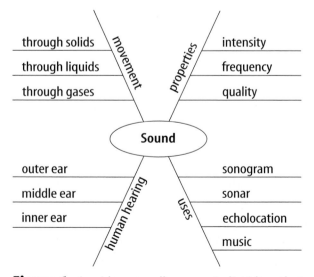

Figure 6 A spider map allows you to list ideas that relate to a central topic but not necessarily to one another.

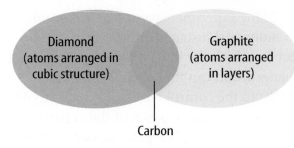

Figure 7 This Venn diagram compares and contrasts two substances made from carbon.

Venn Diagram To illustrate how two subjects compare and contrast you can use a Venn diagram. You can see the characteristics that the subjects have in common and those that they do not, shown in **Figure 7.**

To create a Venn diagram, draw two overlapping ovals that that are big enough to write in. List the characteristics unique to one subject in one oval, and the characteristics of the other subject in the other oval. The characteristics in common are listed in the overlapping section.

Make and Use Tables One way to organize information so it is easier to understand is to use a table. Tables can contain numbers, words, or both.

To make a table, list the items to be compared in the first column and the characteristics to be compared in the first row. The title should clearly indicate the content of the table, and the column or row heads should be clear. Notice that in **Table 1** the units are included.

Table 1 Recyclables Collected During Week			
Day of Week	**Paper (kg)**	**Aluminum (kg)**	**Glass (kg)**
Monday	5.0	4.0	12.0
Wednesday	4.0	1.0	10.0
Friday	2.5	2.0	10.0

Make a Model One way to help you better understand the parts of a structure, the way a process works, or to show things too large or small for viewing is to make a model. For example, an atomic model made of a plastic-ball nucleus and pipe-cleaner electron shells can help you visualize how the parts of an atom relate to each other. Other types of models can by devised on a computer or represented by equations.

Form a Hypothesis

A possible explanation based on previous knowledge and observations is called a hypothesis. After researching gasoline types and recalling previous experiences in your family's car you form a hypothesis—our car runs more efficiently because we use premium gasoline. To be valid, a hypothesis has to be something you can test by using an investigation.

Predict When you apply a hypothesis to a specific situation, you predict something about that situation. A prediction makes a statement in advance, based on prior observation, experience, or scientific reasoning. People use predictions to make everyday decisions. Scientists test predictions by performing investigations. Based on previous observations and experiences, you might form a prediction that cars are more efficient with premium gasoline. The prediction can be tested in an investigation.

Design an Experiment A scientist needs to make many decisions before beginning an investigation. Some of these include: how to carry out the investigation, what steps to follow, how to record the data, and how the investigation will answer the question. It also is important to address any safety concerns.

Test the Hypothesis

Now that you have formed your hypothesis, you need to test it. Using an investigation, you will make observations and collect data, or information. This data might either support or not support your hypothesis. Scientists collect and organize data as numbers and descriptions.

Follow a Procedure In order to know what materials to use, as well as how and in what order to use them, you must follow a procedure. **Figure 8** shows a procedure you might follow to test your hypothesis.

Procedure
1. Use regular gasoline for two weeks.
2. Record the number of kilometers between fill-ups and the amount of gasoline used.
3. Switch to premium gasoline for two weeks.
4. Record the number of kilometers between fill-ups and the amount of gasoline used.

Figure 8 A procedure tells you what to do step by step.

Identify and Manipulate Variables and Controls In any experiment, it is important to keep everything the same except for the item you are testing. The one factor you change is called the independent variable. The change that results is the dependent variable. Make sure you have only one independent variable, to assure yourself of the cause of the changes you observe in the dependent variable. For example, in your gasoline experiment the type of fuel is the independent variable. The dependent variable is the efficiency.

Many experiments also have a control—an individual instance or experimental subject for which the independent variable is not changed. You can then compare the test results to the control results. To design a control you can have two cars of the same type. The control car uses regular gasoline for four weeks. After you are done with the test, you can compare the experimental results to the control results.

Collect Data

Whether you are carrying out an investigation or a short observational experiment, you will collect data, as shown in **Figure 9.** Scientists collect data as numbers and descriptions and organize it in specific ways.

Observe Scientists observe items and events, then record what they see. When they use only words to describe an observation, it is called qualitative data. Scientists' observations also can describe how much there is of something. These observations use numbers, as well as words, in the description and are called quantitative data. For example, if a sample of the element gold is described as being "shiny and very dense" the data are qualitative. Quantitative data on this sample of gold might include "a mass of 30 g and a density of 19.3 g/cm^3."

Figure 9 Collecting data is one way to gather information directly.

Figure 10 Record data neatly and clearly so it is easy to understand.

When you make observations you should examine the entire object or situation first, and then look carefully for details. It is important to record observations accurately and completely. Always record your notes immediately as you make them, so you do not miss details or make a mistake when recording results from memory. Never put unidentified observations on scraps of paper. Instead they should be recorded in a notebook, like the one in **Figure 10.** Write your data neatly so you can easily read it later. At each point in the experiment, record your observations and label them. That way, you will not have to determine what the figures mean when you look at your notes later. Set up any tables that you will need to use ahead of time, so you can record any observations right away. Remember to avoid bias when collecting data by not including personal thoughts when you record observations. Record only what you observe.

Estimate Scientific work also involves estimating. To estimate is to make a judgment about the size or the number of something without measuring or counting. This is important when the number or size of an object or population is too large or too difficult to accurately count or measure.

Sample Scientists may use a sample or a portion of the total number as a type of estimation. To sample is to take a small, representative portion of the objects or organisms of a population for research. By making careful observations or manipulating variables within that portion of the group, information is discovered and conclusions are drawn that might apply to the whole population. A poorly chosen sample can be unrepresentative of the whole. If you were trying to determine the rainfall in an area, it would not be best to take a rainfall sample from under a tree.

Measure You use measurements everyday. Scientists also take measurements when collecting data. When taking measurements, it is important to know how to use measuring tools properly. Accuracy also is important.

Length To measure length, the distance between two points, scientists use meters. Smaller measurements might be measured in centimeters or millimeters.

Length is measured using a metric ruler or meter stick. When using a metric ruler, line up the 0-cm mark with the end of the object being measured and read the number of the unit where the object ends. Look at the metric ruler shown in **Figure 11.** The centimeter lines are the long, numbered lines, and the shorter lines are millimeter lines. In this instance, the length would be 4.50 cm.

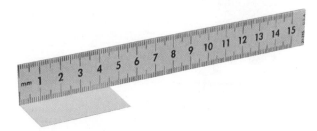

Figure 11 This metric ruler has centimeter and millimeter divisions.

Mass The SI unit for mass is the kilogram (kg). Scientists can measure mass using units formed by adding metric prefixes to the unit gram (g), such as milligram (mg). To measure mass, you might use a triple-beam balance similar to the one shown in **Figure 12.** The balance has a pan on one side and a set of beams on the other side. Each beam has a rider that slides on the beam.

When using a triple-beam balance, place an object on the pan. Slide the largest rider along its beam until the pointer drops below zero. Then move it back one notch. Repeat the process for each rider proceeding from the larger to smaller until the pointer swings an equal distance above and below the zero point. Sum the masses on each beam to find the mass of the object. Move all riders back to zero when finished.

Instead of putting materials directly on the balance, scientists often take a tare of a container. A tare is the mass of a container into which objects or substances are placed for measuring their masses. To mass objects or substances, find the mass of a clean container. Remove the container from the pan, and place the object or substances in the container. Find the mass of the container with the materials in it. Subtract the mass of the empty container from the mass of the filled container to find the mass of the materials you are using.

Figure 12 A triple-beam balance is used to determine the mass of an object.

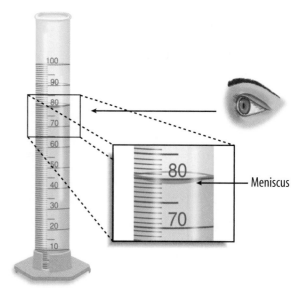

Meniscus

Figure 13 Graduated cylinders measure liquid volume.

Liquid Volume To measure liquids, the unit used is the liter. When a smaller unit is needed, scientists might use a milliliter. Because a milliliter takes up the volume of a cube measuring 1 cm on each side it also can be called a cubic centimeter ($cm^3 = cm \times cm \times cm$).

You can use beakers and graduated cylinders to measure liquid volume. A graduated cylinder, shown in **Figure 13,** is marked from bottom to top in milliliters. In lab, you might use a 10-mL graduated cylinder or a 100-mL graduated cylinder. When measuring liquids, notice that the liquid has a curved surface. Look at the surface at eye level, and measure the bottom of the curve. This is called the meniscus. The graduated cylinder in **Figure 13** contains 79.0 mL, or 79.0 cm^3, of a liquid.

Temperature Scientists often measure temperature using the Celsius scale. Pure water has a freezing point of 0°C and boiling point of 100°C. The unit of measurement is degrees Celsius. Two other scales often used are the Fahrenheit and Kelvin scales.

Analyze the Data

To determine the meaning of your observations and investigation results, you will need to look for patterns in the data. Then you must think critically to determine what the data mean. Scientists use several approaches when they analyze the data they have collected and recorded. Each approach is useful for identifying specific patterns.

Interpret Data The word *interpret* means "to explain the meaning of something." When analyzing data from an experiment, try to find out what the data show. Identify the control group and the test group to see whether or not changes in the independent variable have had an effect. Look for differences in the dependent variable between the control and test groups.

Classify Sorting objects or events into groups based on common features is called classifying. When classifying, first observe the objects or events to be classified. Then select one feature that is shared by some members in the group, but not by all. Place those members that share that feature in a subgroup. You can classify members into smaller and smaller subgroups based on characteristics. Remember that when you classify, you are grouping objects or events for a purpose. Keep your purpose in mind as you select the features to form groups and subgroups.

Compare and Contrast Observations can be analyzed by noting the similarities and differences between two more objects or events that you observe. When you look at objects or events to see how they are similar, you are comparing them. Contrasting is looking for differences in objects or events.

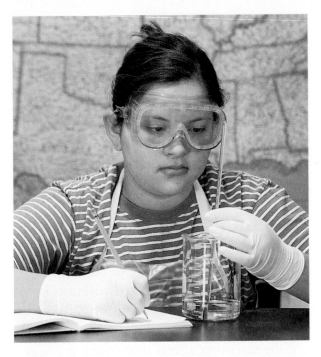

Figure 14 A thermometer measures the temperature of an object.

Scientists use a thermometer to measure temperature. Most thermometers in a laboratory are glass tubes with a bulb at the bottom end containing a liquid such as colored alcohol. The liquid rises or falls with a change in temperature. To read a glass thermometer like the thermometer in **Figure 14,** rotate it slowly until a red line appears. Read the temperature where the red line ends.

Form Operational Definitions An operational definition defines an object by how it functions, works, or behaves. For example, when you are playing hide and seek and a tree is home base, you have created an operational definition for a tree.

Objects can have more than one operational definition. For example, a ruler can be defined as a tool that measures the length of an object (how it is used). It can also be a tool with a series of marks used as a standard when measuring (how it works).

Recognize Cause and Effect A cause is a reason for an action or condition. The effect is that action or condition. When two events happen together, it is not necessarily true that one event caused the other. Scientists must design a controlled investigation to recognize the exact cause and effect.

Draw Conclusions

When scientists have analyzed the data they collected, they proceed to draw conclusions about the data. These conclusions are sometimes stated in words similar to the hypothesis that you formed earlier. They may confirm a hypothesis, or lead you to a new hypothesis.

Infer Scientists often make inferences based on their observations. An inference is an attempt to explain observations or to indicate a cause. An inference is not a fact, but a logical conclusion that needs further investigation. For example, you may infer that a fire has caused smoke. Until you investigate, however, you do not know for sure.

Apply When you draw a conclusion, you must apply those conclusions to determine whether the data supports the hypothesis. If your data do not support your hypothesis, it does not mean that the hypothesis is wrong. It means only that the result of the investigation did not support the hypothesis. Maybe the experiment needs to be redesigned, or some of the initial observations on which the hypothesis was based were incomplete or biased. Perhaps more observation or research is needed to refine your hypothesis. A successful investigation does not always come out the way you originally predicted.

Avoid Bias Sometimes a scientific investigation involves making judgments. When you make a judgment, you form an opinion. It is important to be honest and not to allow any expectations of results to bias your judgments. This is important throughout the entire investigation, from researching to collecting data to drawing conclusions.

Communicate

The communication of ideas is an important part of the work of scientists. A discovery that is not reported will not advance the scientific community's understanding or knowledge. Communication among scientists also is important as a way of improving their investigations.

Scientists communicate in many ways, from writing articles in journals and magazines that explain their investigations and experiments, to announcing important discoveries on television and radio. Scientists also share ideas with colleagues on the Internet or present them as lectures, like the student is doing in **Figure 15.**

Figure 15 A student communicates to his peers about his investigation.

SAFETY SYMBOLS

SAFETY SYMBOLS	HAZARD	EXAMPLES	PRECAUTION	REMEDY
DISPOSAL	Special disposal procedures need to be followed.	certain chemicals, living organisms	Do not dispose of these materials in the sink or trash can.	Dispose of wastes as directed by your teacher.
BIOLOGICAL	Organisms or other biological materials that might be harmful to humans	bacteria, fungi, blood, unpreserved tissues, plant materials	Avoid skin contact with these materials. Wear mask or gloves.	Notify your teacher if you suspect contact with material. Wash hands thoroughly.
EXTREME TEMPERATURE	Objects that can burn skin by being too cold or too hot	boiling liquids, hot plates, dry ice, liquid nitrogen	Use proper protection when handling.	Go to your teacher for first aid.
SHARP OBJECT	Use of tools or glassware that can easily puncture or slice skin	razor blades, pins, scalpels, pointed tools, dissecting probes, broken glass	Practice common-sense behavior and follow guidelines for use of the tool.	Go to your teacher for first aid.
FUME	Possible danger to respiratory tract from fumes	ammonia, acetone, nail polish remover, heated sulfur, moth balls	Make sure there is good ventilation. Never smell fumes directly. Wear a mask.	Leave foul area and notify your teacher immediately.
ELECTRICAL	Possible danger from electrical shock or burn	improper grounding, liquid spills, short circuits, exposed wires	Double-check setup with teacher. Check condition of wires and apparatus.	Do not attempt to fix electrical problems. Notify your teacher immediately.
IRRITANT	Substances that can irritate the skin or mucous membranes of the respiratory tract	pollen, moth balls, steel wool, fiberglass, potassium permanganate	Wear dust mask and gloves. Practice extra care when handling these materials.	Go to your teacher for first aid.
CHEMICAL	Chemicals can react with and destroy tissue and other materials	bleaches such as hydrogen peroxide; acids such as sulfuric acid, hydrochloric acid; bases such as ammonia, sodium hydroxide	Wear goggles, gloves, and an apron.	Immediately flush the affected area with water and notify your teacher.
TOXIC	Substance may be poisonous if touched, inhaled, or swallowed.	mercury, many metal compounds, iodine, poinsettia plant parts	Follow your teacher's instructions.	Always wash hands thoroughly after use. Go to your teacher for first aid.
FLAMMABLE	Flammable chemicals may be ignited by open flame, spark, or exposed heat.	alcohol, kerosene, potassium permanganate	Avoid open flames and heat when using flammable chemicals.	Notify your teacher immediately. Use fire safety equipment if applicable.
OPEN FLAME	Open flame in use, may cause fire.	hair, clothing, paper, synthetic materials	Tie back hair and loose clothing. Follow teacher's instruction on lighting and extinguishing flames.	Notify your teacher immediately. Use fire safety equipment if applicable.

 Eye Safety
Proper eye protection should be worn at all times by anyone performing or observing science activities.

 Clothing Protection
This symbol appears when substances could stain or burn clothing.

 Animal Safety
This symbol appears when safety of animals and students must be ensured.

 Handwashing
After the lab, wash hands with soap and water before removing goggles.

Safety in the Science Laboratory

The science laboratory is a safe place to work if you follow standard safety procedures. Being responsible for your own safety helps to make the entire laboratory a safer place for everyone. When performing any lab, read and apply the caution statements and safety symbol listed at the beginning of the lab.

General Safety Rules

1. Obtain your teacher's permission to begin all investigations and use laboratory equipment.

2. Study the procedure. Ask your teacher any questions. Be sure you understand safety symbols shown on the page.

3. Notify your teacher about allergies or other health conditions which can affect your participation in a lab.

4. Learn and follow use and safety procedures for your equipment. If unsure, ask your teacher.

5. Never eat, drink, chew gum, apply cosmetics, or do any personal grooming in the lab. Never use lab glassware as food or drink containers. Keep your hands away from your face and mouth.

6. Know the location and proper use of the safety shower, eye wash, fire blanket, and fire alarm.

Prevent Accidents

1. Use the safety equipment provided to you. Goggles and a safety apron should be worn during investigations.

2. Do NOT use hair spray, mousse, or other flammable hair products. Tie back long hair and tie down loose clothing.

3. Do NOT wear sandals or other open-toed shoes in the lab.

4. Remove jewelry on hands and wrists. Loose jewelry, such as chains and long necklaces, should be removed to prevent them from getting caught in equipment.

5. Do not taste any substances or draw any material into a tube with your mouth.

6. Proper behavior is expected in the lab. Practical jokes and fooling around can lead to accidents and injury.

7. Keep your work area uncluttered.

Laboratory Work

1. Collect and carry all equipment and materials to your work area before beginning a lab.

2. Remain in your own work area unless given permission by your teacher to leave it.

3. Always slant test tubes away from yourself and others when heating them, adding substances to them, or rinsing them.

4. If instructed to smell a substance in a container, hold the container a short distance away and fan vapors towards your nose.

5. Do NOT substitute other chemicals/substances for those in the materials list unless instructed to do so by your teacher.

6. Do NOT take any materials or chemicals outside of the laboratory.

7. Stay out of storage areas unless instructed to be there and supervised by your teacher.

Laboratory Cleanup

1. Turn off all burners, water, and gas, and disconnect all electrical devices.

2. Clean all pieces of equipment and return all materials to their proper places.

3. Dispose of chemicals and other materials as directed by your teacher. Place broken glass and solid substances in the proper containers. Never discard materials in the sink.

4. Clean your work area.

5. Wash your hands with soap and water thoroughly BEFORE removing your goggles.

Emergencies

1. Report any fire, electrical shock, glassware breakage, spill, or injury, no matter how small, to your teacher immediately. Follow his or her instructions.

2. If your clothing should catch fire, STOP, DROP, and ROLL. If possible, smother it with the fire blanket or get under a safety shower. NEVER RUN.

3. If a fire should occur, turn off all gas and leave the room according to established procedures.

4. In most instances, your teacher will clean up spills. Do NOT attempt to clean up spills unless you are given permission and instructions to do so.

5. If chemicals come into contact with your eyes or skin, notify your teacher immediately. Use the eyewash or flush your skin or eyes with large quantities of water.

6. The fire extinguisher and first-aid kit should only be used by your teacher unless it is an extreme emergency and you have been given permission.

7. If someone is injured or becomes ill, only a professional medical provider or someone certified in first aid should perform first-aid procedures.

EXTRA  Labs

From Your Kitchen, Junk Drawer, or Yard

1 Rock Creatures

▶ *Real-World Question*

What types of organisms live under stream rocks?

Possible Materials
- waterproof boots
- ice cube tray (white)
- aquarium net
- bucket
- collecting jars
- guidebook to pond life

▶ *Procedure*

1. With permission, search under the rocks of a local stream. Look for aquatic organisms under the rocks and leaves of the stream. Compare what you find in fast- and slow-moving water.

2. With permission, carefully pull organisms you find off the rocks and put them into separate compartments of your ice cube tray. Take care not to injure the creatures you find.

3. Use your net and bucket to collect larger organisms.

4. Use your guidebook to pond life to identify the organisms you find.

5. Release the organisms back into the stream once you identify them.

▶ *Conclude and Apply*

1. Identify and list the organisms you found under the stream rocks.

2. Infer why so many aquatic organisms make their habitats beneath stream rocks.

2 A Light in the Forest

▶ *Real-World Question*

Does the amount of sunlight vary in a forest?

Possible Materials
- empty toilet paper or paper towel roll
- Science Journal

▶ *Procedure:*

1. Copy the data table into your Science Journal.

2. Go with an adult to a nearby forest or large grove of trees.

3. Stand near the edge of the forest and look straight up through your cardboard tube. Estimate the percentage of blue sky and clouds you can see in the circle. This percentage is the amount of sunlight reaching the forest floor.

4. Record your location and estimated percentage of sunlight in your data table.

5. Test several other locations in the forest. Choose places where the trees completely cover the forest floor and where sunlight is partially coming through.

Data Table

Location	% of Sunlight

▶ *Conclude and Apply*

1. Explain how the amount of sunlight reaching the forest floor changed.

2. Infer why it is important for leaves and branches to stop sunlight from reaching much of the forest floor.

Adult supervision required for all labs.

3 Echinoderm Hold

Real-World Question
How do echinoderms living in intertidal ecosystems hold on to rocks?

Possible Materials
- plastic suction cup
- water
- paper towel or sponge

Procedure
1. Moisten a paper towel or sponge with water.
2. Press a plastic suction cup on the moist towel or sponge until the entire bottom surface of the cup is wet.
3. Firmly press the suction cup down on a kitchen counter for 10 s.
4. Grab the top handle of the suction cup and try removing the cup from the counter by pulling it straight up.

Conclude and Apply
1. Describe what happened when you tried to remove the cup from the counter.
2. Infer how echinoderms living in intertidal ecosystems withstand the constant pull of ocean waves and currents.

4 UV Watch

Real-World Question
How can you find out about the risks of ultraviolet radiation each day?

Possible Materials
- daily newspaper with weekly weather forecasts
- graph paper

Procedure
1. Use the local newspaper or another resource to get the weather forecast for the day.
2. Check the UV (ultraviolet light) index for the day. If it provides an hourly UV index level, record the level for 1:00 P.M.
3. Find a legend or do research to discover what the numbers of the UV index mean.
4. Record the UV index everyday for ten days and graph your results on graph paper.

Conclude and Apply
1. Explain how the UV index system works.
2. Research several ways you can protect yourself from too much ultraviolet light exposure.

⑤ Biodiversity

▶ Real-World Question

How does biodiversity vary around your home?

Possible Materials 🥽 📷 ✂️ 📐
- wooden ground stakes or nails
- string
- poster paper
- measuring tape, yard stick, or ruler

▶ Procedure

1. With an adult, visit a wild area near your home or school. Mark off an area that is 1 m by 1 m. Put stakes at the four corners and string between them to make a fence around your plot.
2. Count or estimate how many of each animal or plant species are in your plot.

3. Make a drawing on your poster paper and a list of the species found. If you cannot identify a species by name, just describe and sketch it.
4. Repeat the procedure on a lawn. Make your sketch and list on the opposite side of the poster paper.

▶ Conclude and Apply

1. Describe the difference in biodiversity between the lawn plot and the wild plot.
2. Think about a prize flower garden. What is the biodiversity like in the flower garden, compared to the wild?
3. Would you find more bird species in the wild, or near the lawn or flower garden? Why?

Adult supervision required for all labs.

Computer Skills

People who study science rely on computers, like the one in **Figure 16,** to record and store data and to analyze results from investigations. Whether you work in a laboratory or just need to write a lab report with tables, good computer skills are a necessity.

Using the computer comes with responsibility. Issues of ownership, security, and privacy can arise. Remember, if you did not author the information you are using, you must provide a source for your information. Also, anything on a computer can be accessed by others. Do not put anything on the computer that you would not want everyone to know. To add more security to your work, use a password.

Use a Word Processing Program

A computer program that allows you to type your information, change it as many times as you need to, and then print it out is called a word processing program. Word processing programs also can be used to make tables.

Figure 16 A computer will make reports neater and more professional looking.

Learn the Skill To start your word processing program, a blank document, sometimes called "Document 1," appears on the screen. To begin, start typing. To create a new document, click the *New* button on the standard tool bar. These tips will help you format the document.

- The program will automatically move to the next line; press *Enter* if you wish to start a new paragraph.
- Symbols, called non-printing characters, can be hidden by clicking the *Show/Hide* button on your toolbar.
- To insert text, move the cursor to the point where you want the insertion to go, click on the mouse once, and type the text.
- To move several lines of text, select the text and click the *Cut* button on your toolbar. Then position your cursor in the location that you want to move the cut text and click *Paste.* If you move to the wrong place, click *Undo.*
- The spell check feature does not catch words that are misspelled to look like other words, like "cold" instead of "gold." Always reread your document to catch all spelling mistakes.
- To learn about other word processing methods, read the user's manual or click on the *Help* button.
- You can integrate databases, graphics, and spreadsheets into documents by copying from another program and pasting it into your document, or by using desktop publishing (DTP). DTP software allows you to put text and graphics together to finish your document with a professional look. This software varies in how it is used and its capabilities.

Use a Database

A collection of facts stored in a computer and sorted into different fields is called a database. A database can be reorganized in any way that suits your needs.

Learn the Skill A computer program that allows you to create your own database is a database management system (DBMS). It allows you to add, delete, or change information. Take time to get to know the features of your database software.

- Determine what facts you would like to include and research to collect your information.
- Determine how you want to organize the information.
- Follow the instructions for your particular DBMS to set up fields. Then enter each item of data in the appropriate field.
- Follow the instructions to sort the information in order of importance.
- Evaluate the information in your database, and add, delete, or change as necessary.

Use the Internet

The Internet is a global network of computers where information is stored and shared. To use the Internet, like the students in **Figure 17,** you need a modem to connect your computer to a phone line and an Internet Service Provider account.

Learn the Skill To access internet sites and information, use a "Web browser," which lets you view and explore pages on the World Wide Web. Each page is its own site, and each site has its own address, called a URL. Once you have found a Web browser, follow these steps for a search (this also is how you search a database).

Figure 17 The Internet allows you to search a global network for a variety of information.

- Be as specific as possible. If you know you want to research "gold," don't type in "elements." Keep narrowing your search until you find what you want.
- Web sites that end in *.com* are commercial Web sites; *.org, .edu,* and *.gov* are non-profit, educational, or government Web sites.
- Electronic encyclopedias, almanacs, indexes, and catalogs will help locate and select relevant information.
- Develop a "home page" with relative ease. When developing a Web site, NEVER post pictures or disclose personal information such as location, names, or phone numbers. Your school or community usually can host your Web site. A basic understanding of HTML (hypertext mark-up language), the language of Web sites, is necessary. Software that creates HTML code is called authoring software, and can be downloaded free from many Web sites. This software allows text and pictures to be arranged as the software is writing the HTML code.

Use a Spreadsheet

A spreadsheet, shown in **Figure 18,** can perform mathematical functions with any data arranged in columns and rows. By entering a simple equation into a cell, the program can perform operations in specific cells, rows, or columns.

Learn the Skill Each column (vertical) is assigned a letter, and each row (horizontal) is assigned a number. Each point where a row and column intersect is called a cell, and is labeled according to where it is located—Column A, Row 1 (A1).

- Decide how to organize the data, and enter it in the correct row or column.
- Spreadsheets can use standard formulas or formulas can be customized to calculate cells.
- To make a change, click on a cell to make it activate, and enter the edited data or formula.
- Spreadsheets also can display your results in graphs. Choose the style of graph that best represents the data.

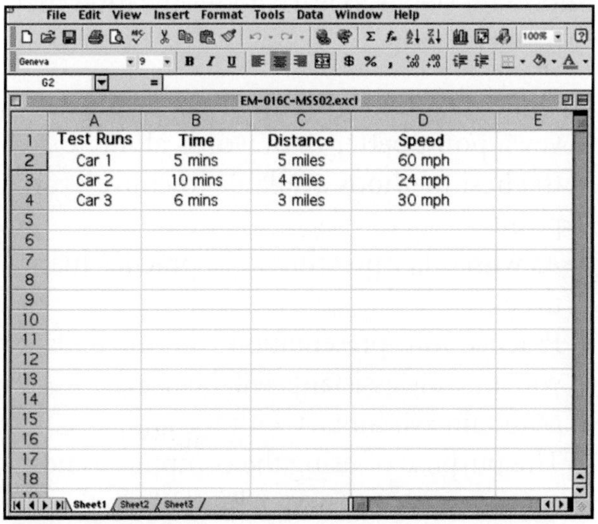

	A	B	C	D	E
1	Test Runs	Time	Distance	Speed	
2	Car 1	5 mins	5 miles	60 mph	
3	Car 2	10 mins	4 miles	24 mph	
4	Car 3	6 mins	3 miles	30 mph	

Figure 18 A spreadsheet allows you to perform mathematical operations on your data.

Use Graphics Software

Adding pictures, called graphics, to your documents is one way to make your documents more meaningful and exciting. This software adds, edits, and even constructs graphics. There is a variety of graphics software programs. The tools used for drawing can be a mouse, keyboard, or other specialized devices. Some graphics programs are simple. Others are complicated, called computer-aided design (CAD) software.

Learn the Skill It is important to have an understanding of the graphics software being used before starting. The better the software is understood, the better the results. The graphics can be placed in a word-processing document.

- Clip art can be found on a variety of internet sites, and on CDs. These images can be copied and pasted into your document.
- When beginning, try editing existing drawings, then work up to creating drawings.
- The images are made of tiny rectangles of color called pixels. Each pixel can be altered.
- Digital photography is another way to add images. The photographs in the memory of a digital camera can be downloaded into a computer, then edited and added to the document.
- Graphics software also can allow animation. The software allows drawings to have the appearance of movement by connecting basic drawings automatically. This is called in-betweening, or tweening.
- Remember to save often.

Presentation Skills

Develop Multimedia Presentations

Most presentations are more dynamic if they include diagrams, photographs, videos, or sound recordings, like the one shown in **Figure 19.** A multimedia presentation involves using stereos, overhead projectors, televisions, computers, and more.

Learn the Skill Decide the main points of your presentation, and what types of media would best illustrate those points.

- Make sure you know how to use the equipment you are working with.
- Practice the presentation using the equipment several times.
- Enlist the help of a classmate to push play or turn lights out for you. Be sure to practice your presentation with him or her.
- If possible, set up all of the equipment ahead of time, and make sure everything is working properly.

Figure 19 These students are engaging the audience using a variety of tools.

Computer Presentations

There are many different interactive computer programs that you can use to enhance your presentation. Most computers have a compact disc (CD) drive that can play both CDs and digital video discs (DVDs). Also, there is hardware to connect a regular CD, DVD, or VCR. These tools will enhance your presentation.

Another method of using the computer to aid in your presentation is to develop a slide show using a computer program. This can allow movement of visuals at the presenter's pace, and can allow for visuals to build on one another.

Learn the Skill In order to create multimedia presentations on a computer, you need to have certain tools. These may include traditional graphic tools and drawing programs, animation programs, and authoring systems that tie everything together. Your computer will tell you which tools it supports. The most important step is to learn about the tools that you will be using.

- Often, color and strong images will convey a point better than words alone. Use the best methods available to convey your point.
- As with other presentations, practice many times.
- Practice your presentation with the tools you and any assistants will be using.
- Maintain eye contact with the audience. The purpose of using the computer is not to prompt the presenter, but to help the audience understand the points of the presentation.

Math Review

Use Fractions

A fraction compares a part to a whole. In the fraction $\frac{2}{3}$, the 2 represents the part and is the numerator. The 3 represents the whole and is the denominator.

Reduce Fractions To reduce a fraction, you must find the largest factor that is common to both the numerator and the denominator, the greatest common factor (GCF). Divide both numbers by the GCF. The fraction has then been reduced, or it is in its simplest form.

Example Twelve of the 20 chemicals in the science lab are in powder form. What fraction of the chemicals used in the lab are in powder form?

Step 1 Write the fraction.

$$\frac{\text{part}}{\text{whole}} = \frac{12}{20}$$

Step 2 To find the GCF of the numerator and denominator, list all of the factors of each number.

Factors of 12: 1, 2, 3, 4, 6, 12 (the numbers that divide evenly into 12)

Factors of 20: 1, 2, 4, 5, 10, 20 (the numbers that divide evenly into 20)

Step 3 List the common factors.

1, 2, 4.

Step 4 Choose the greatest factor in the list.

The GCF of 12 and 20 is 4.

Step 5 Divide the numerator and denominator by the GCF.

$$\frac{12 \div 4}{20 \div 4} = \frac{3}{5}$$

In the lab, $\frac{3}{5}$ of the chemicals are in powder form.

Practice Problem At an amusement park, 66 of 90 rides have a height restriction. What fraction of the rides, in its simplest form, has a height restriction?

Add and Subtract Fractions To add or subtract fractions with the same denominator, add or subtract the numerators and write the sum or difference over the denominator. After finding the sum or difference, find the simplest form for your fraction.

Example 1 In the forest outside your house, $\frac{1}{8}$ of the animals are rabbits, $\frac{3}{8}$ are squirrels, and the remainder are birds and insects. How many are mammals?

Step 1 Add the numerators.

$$\frac{1}{8} + \frac{3}{8} = \frac{(1 + 3)}{8} = \frac{4}{8}$$

Step 2 Find the GCF.

$$\frac{4}{8} \quad (\text{GCF, } 4)$$

Step 3 Divide the numerator and denominator by the GCF.

$$\frac{4}{4} = 1, \quad \frac{8}{4} = 2$$

$\frac{1}{2}$ of the animals are mammals.

Example 2 If $\frac{7}{16}$ of the Earth is covered by freshwater, and $\frac{1}{16}$ of that is in glaciers, how much freshwater is not frozen?

Step 1 Subtract the numerators.

$$\frac{7}{16} - \frac{1}{16} = \frac{(7 - 1)}{16} = \frac{6}{16}$$

Step 2 Find the GCF.

$$\frac{6}{16} \quad (\text{GCF, } 2)$$

Step 3 Divide the numerator and denominator by the GCF.

$$\frac{6}{2} = 3, \quad \frac{16}{2} = 8$$

$\frac{3}{8}$ of the freshwater is not frozen.

Practice Problem A bicycle rider is going 15 km/h for $\frac{4}{9}$ of his ride, 10 km/h for $\frac{2}{9}$ of his ride, and 8 km/h for the remainder of the ride. How much of his ride is he going over 8 km/h?

Unlike Denominators To add or subtract fractions with unlike denominators, first find the least common denominator (LCD). This is the smallest number that is a common multiple of both denominators. Rename each fraction with the LCD, and then add or subtract. Find the simplest form if necessary.

Example 1 A chemist makes a paste that is $\frac{1}{2}$ table salt (NaCl), $\frac{1}{3}$ sugar ($C_6H_{12}O_6$), and the rest water (H_2O). How much of the paste is a solid?

Step 1 Find the LCD of the fractions.

$\frac{1}{2} + \frac{1}{3}$ (LCD, 6)

Step 2 Rename each numerator and each denominator with the LCD.

$1 \times 3 = 3, \ 2 \times 3 = 6$

$1 \times 2 = 2, \ 3 \times 2 = 6$

Step 3 Add the numerators.

$\frac{3}{6} + \frac{2}{6} = \frac{(3 + 2)}{6} = \frac{5}{6}$

$\frac{5}{6}$ of the paste is a solid.

Example 2 The average precipitation in Grand Junction, CO, is $\frac{7}{10}$ inch in November, and $\frac{3}{5}$ inch in December. What is the total average precipitation?

Step 1 Find the LCD of the fractions.

$\frac{7}{10} + \frac{3}{5}$ (LCD, 10)

Step 2 Rename each numerator and each denominator with the LCD.

$7 \times 1 = 7, \ 10 \times 1 = 10$

$3 \times 2 = 6, \ 5 \times 2 = 10$

Step 3 Add the numerators.

$\frac{7}{10} + \frac{6}{10} = \frac{(7 + 6)}{10} = \frac{13}{10}$

$\frac{13}{10}$ inches total precipitation, or $1\frac{3}{10}$ inches.

Practice Problem On an electric bill, about $\frac{1}{8}$ of the energy is from solar energy and about $\frac{1}{10}$ is from wind power. How much of the total bill is from solar energy and wind power combined?

Example 3 In your body, $\frac{7}{10}$ of your muscle contractions are involuntary (cardiac and smooth muscle tissue). Smooth muscle makes $\frac{3}{15}$ of your muscle contractions. How many of your muscle contractions are made by cardiac muscle?

Step 1 Find the LCD of the fractions.

$\frac{7}{10} - \frac{3}{15}$ (LCD, 30)

Step 2 Rename each numerator and each denominator with the LCD.

$7 \times 3 = 21, \ 10 \times 3 = 30$

$3 \times 2 = 6, \ 15 \times 2 = 30$

Step 3 Subtract the numerators.

$\frac{21}{30} - \frac{6}{30} = \frac{(21 - 6)}{30} = \frac{15}{30}$

Step 4 Find the GCF.

$\frac{15}{30}$ (GCF, 15)

$\frac{1}{2}$

$\frac{1}{2}$ of all muscle contractions are cardiac muscle.

Example 4 Tony wants to make cookies that call for $\frac{3}{4}$ of a cup of flour, but he only has $\frac{1}{3}$ of a cup. How much more flour does he need?

Step 1 Find the LCD of the fractions.

$\frac{3}{4} - \frac{1}{3}$ (LCD, 12)

Step 2 Rename each numerator and each denominator with the LCD.

$3 \times 3 = 9, \ 4 \times 3 = 12$

$1 \times 4 = 4, \ 3 \times 4 = 12$

Step 3 Subtract the numerators.

$\frac{9}{12} - \frac{4}{12} = \frac{(9 - 4)}{12} = \frac{5}{12}$

$\frac{5}{12}$ of a cup of flour.

Practice Problem Using the information provided to you in Example 3 above, determine how many muscle contractions are voluntary (skeletal muscle).

Multiply Fractions To multiply with fractions, multiply the numerators and multiply the denominators. Find the simplest form if necessary.

Example Multiply $\frac{3}{5}$ by $\frac{1}{3}$.

Step 1 Multiply the numerators and denominators.

$$\frac{3}{5} \times \frac{1}{3} = \frac{(3 \times 1)}{(5 \times 3)} = \frac{3}{15}$$

Step 2 Find the GCF.

$$\frac{3}{15} \quad (GCF, 3)$$

Step 3 Divide the numerator and denominator by the GCF.

$$\frac{3}{3} = 1, \ \frac{15}{3} = 5$$

$$\frac{1}{5}$$

$\frac{3}{5}$ multiplied by $\frac{1}{3}$ is $\frac{1}{5}$.

Practice Problem Multiply $\frac{3}{14}$ by $\frac{5}{16}$.

Find a Reciprocal Two numbers whose product is 1 are called multiplicative inverses, or reciprocals.

Example Find the reciprocal of $\frac{3}{8}$.

Step 1 Inverse the fraction by putting the denominator on top and the numerator on the bottom.

$$\frac{8}{3}$$

The reciprocal of $\frac{3}{8}$ is $\frac{8}{3}$.

Practice Problem Find the reciprocal of $\frac{4}{9}$.

Divide Fractions To divide one fraction by another fraction, multiply the dividend by the reciprocal of the divisor. Find the simplest form if necessary.

Example 1 Divide $\frac{1}{9}$ by $\frac{1}{3}$.

Step 1 Find the reciprocal of the divisor.

The reciprocal of $\frac{1}{3}$ is $\frac{3}{1}$.

Step 2 Multiply the dividend by the reciprocal of the divisor.

$$\frac{\frac{1}{9}}{\frac{1}{3}} = \frac{1}{9} \times \frac{3}{1} = \frac{(1 \times 3)}{(9 \times 1)} = \frac{3}{9}$$

Step 3 Find the GCF.

$$\frac{3}{9} \quad (GCF, 3)$$

Step 4 Divide the numerator and denominator by the GCF.

$$\frac{3}{3} = 1, \ \frac{9}{3} = 3$$

$$\frac{1}{3}$$

$\frac{1}{9}$ divided by $\frac{1}{3}$ is $\frac{1}{3}$.

Example 2 Divide $\frac{3}{5}$ by $\frac{1}{4}$.

Step 1 Find the reciprocal of the divisor.

The reciprocal of $\frac{1}{4}$ is $\frac{4}{1}$.

Step 2 Multiply the dividend by the reciprocal of the divisor.

$$\frac{\frac{3}{5}}{\frac{1}{4}} = \frac{3}{5} \times \frac{4}{1} = \frac{(3 \times 4)}{(5 \times 1)} = \frac{12}{5}$$

$\frac{3}{5}$ divided by $\frac{1}{4}$ is $\frac{12}{5}$ or $2\frac{2}{5}$.

Practice Problem Divide $\frac{3}{11}$ by $\frac{7}{10}$.

Use Ratios

When you compare two numbers by division, you are using a ratio. Ratios can be written 3 to 5, 3:5, or $\frac{3}{5}$. Ratios, like fractions, also can be written in simplest form.

Ratios can represent probabilities, also called odds. This is a ratio that compares the number of ways a certain outcome occurs to the number of outcomes. For example, if you flip a coin 100 times, what are the odds that it will come up heads? There are two possible outcomes, heads or tails, so the odds of coming up heads are 50:100. Another way to say this is that 50 out of 100 times the coin will come up heads. In its simplest form, the ratio is 1:2.

Example 1 A chemical solution contains 40 g of salt and 64 g of baking soda. What is the ratio of salt to baking soda as a fraction in simplest form?

Step 1 Write the ratio as a fraction.
$$\frac{\text{salt}}{\text{baking soda}} = \frac{40}{64}$$

Step 2 Express the fraction in simplest form.
The GCF of 40 and 64 is 8.
$$\frac{40}{64} = \frac{40 \div 8}{64 \div 8} = \frac{5}{8}$$

The ratio of salt to baking soda in the sample is 5:8.

Example 2 Sean rolls a 6-sided die 6 times. What are the odds that the side with a 3 will show?

Step 1 Write the ratio as a fraction.
$$\frac{\text{number of sides with a 3}}{\text{number of sides}} = \frac{1}{6}$$

Step 2 Multiply by the number of attempts.
$$\frac{1}{6} \times 6 \text{ attempts} = \frac{6}{6} \text{ attempts} = 1 \text{ attempt}$$

1 attempt out of 6 will show a 3.

Practice Problem Two metal rods measure 100 cm and 144 cm in length. What is the ratio of their lengths in simplest form?

Use Decimals

A fraction with a denominator that is a power of ten can be written as a decimal. For example, 0.27 means $\frac{27}{100}$. The decimal point separates the ones place from the tenths place.

Any fraction can be written as a decimal using division. For example, the fraction $\frac{5}{8}$ can be written as a decimal by dividing 5 by 8. Written as a decimal, it is 0.625.

Add or Subtract Decimals When adding and subtracting decimals, line up the decimal points before carrying out the operation.

Example 1 Find the sum of 47.68 and 7.80.

Step 1 Line up the decimal places when you write the numbers.
```
  47.68
+  7.80
```

Step 2 Add the decimals.
```
  47.68
+  7.80
-------
  55.48
```

The sum of 47.68 and 7.80 is 55.48.

Example 2 Find the difference of 42.17 and 15.85.

Step 1 Line up the decimal places when you write the number.
```
  42.17
- 15.85
```

Step 2 Subtract the decimals.
```
  42.17
- 15.85
-------
  26.32
```

The difference of 42.17 and 15.85 is 26.32.

Practice Problem Find the sum of 1.245 and 3.842.

Multiply Decimals To multiply decimals, multiply the numbers like any other number, ignoring the decimal point. Count the decimal places in each factor. The product will have the same number of decimal places as the sum of the decimal places in the factors.

Example Multiply 2.4 by 5.9.

Step 1 Multiply the factors like two whole numbers.
$24 \times 59 = 1416$

Step 2 Find the sum of the number of decimal places in the factors. Each factor has one decimal place, for a sum of two decimal places.

Step 3 The product will have two decimal places.
14.16

The product of 2.4 and 5.9 is 14.16.

Practice Problem Multiply 4.6 by 2.2.

Divide Decimals When dividing decimals, change the divisor to a whole number. To do this, multiply both the divisor and the dividend by the same power of ten. Then place the decimal point in the quotient directly above the decimal point in the dividend. Then divide as you do with whole numbers.

Example Divide 8.84 by 3.4.

Step 1 Multiply both factors by 10.
$3.4 \times 10 = 34, 8.84 \times 10 = 88.4$

Step 2 Divide 88.4 by 34.

$$
\begin{array}{r}
2.6 \\
34\overline{)88.4} \\
-68 \\
\hline
204 \\
-204 \\
\hline
0
\end{array}
$$

8.84 divided by 3.4 is 2.6.

Practice Problem Divide 75.6 by 3.6.

Use Proportions

An equation that shows that two ratios are equivalent is a proportion. The ratios $\frac{2}{4}$ and $\frac{5}{10}$ are equivalent, so they can be written as $\frac{2}{4} = \frac{5}{10}$. This equation is a proportion.

When two ratios form a proportion, the cross products are equal. To find the cross products in the proportion $\frac{2}{4} = \frac{5}{10}$, multiply the 2 and the 10, and the 4 and the 5. Therefore $2 \times 10 = 4 \times 5$, or $20 = 20$.

Because you know that both proportions are equal, you can use cross products to find a missing term in a proportion. This is known as solving the proportion.

Example The heights of a tree and a pole are proportional to the lengths of their shadows. The tree casts a shadow of 24 m when a 6-m pole casts a shadow of 4 m. What is the height of the tree?

Step 1 Write a proportion.
$$\frac{\text{height of tree}}{\text{height of pole}} = \frac{\text{length of tree's shadow}}{\text{length of pole's shadow}}$$

Step 2 Substitute the known values into the proportion. Let h represent the unknown value, the height of the tree.
$$\frac{h}{6} = \frac{24}{4}$$

Step 3 Find the cross products.
$$h \times 4 = 6 \times 24$$

Step 4 Simplify the equation.
$$4h = 144$$

Step 5 Divide each side by 4.
$$\frac{4h}{4} = \frac{144}{4}$$
$$h = 36$$

The height of the tree is 36 m.

Practice Problem The ratios of the weights of two objects on the Moon and on Earth are in proportion. A rock weighing 3 N on the Moon weighs 18 N on Earth. How much would a rock that weighs 5 N on the Moon weigh on Earth?

Use Percentages

The word *percent* means "out of one hundred." It is a ratio that compares a number to 100. Suppose you read that 77 percent of the Earth's surface is covered by water. That is the same as reading that the fraction of the Earth's surface covered by water is $\frac{77}{100}$. To express a fraction as a percent, first find the equivalent decimal for the fraction. Then, multiply the decimal by 100 and add the percent symbol.

Example Express $\frac{13}{20}$ as a percent.

Step 1 Find the equivalent decimal for the fraction.

$$\begin{array}{r} 0.65 \\ 20\overline{)13.00} \\ \underline{12\ 0} \\ 1\ 00 \\ \underline{1\ 00} \\ 0 \end{array}$$

Step 2 Rewrite the fraction $\frac{13}{20}$ as 0.65.

Step 3 Multiply 0.65 by 100 and add the % sign.
$$0.65 \times 100 = 65 = 65\%$$

So, $\frac{13}{20} = 65\%$.

This also can be solved as a proportion.

Example Express $\frac{13}{20}$ as a percent.

Step 1 Write a proportion.
$$\frac{13}{20} = \frac{x}{100}$$

Step 2 Find the cross products.
$$1300 = 20x$$

Step 3 Divide each side by 20.
$$\frac{1300}{20} = \frac{20x}{20}$$
$$65\% = x$$

Practice Problem In one year, 73 of 365 days were rainy in one city. What percent of the days in that city were rainy?

Solve One-Step Equations

A statement that two things are equal is an equation. For example, $A = B$ is an equation that states that A is equal to B.

An equation is solved when a variable is replaced with a value that makes both sides of the equation equal. To make both sides equal the inverse operation is used. Addition and subtraction are inverses, and multiplication and division are inverses.

Example 1 Solve the equation $x - 10 = 35$.

Step 1 Find the solution by adding 10 to each side of the equation.
$$x - 10 = 35$$
$$x - 10 + 10 = 35 + 10$$
$$x = 45$$

Step 2 Check the solution.
$$x - 10 = 35$$
$$45 - 10 = 35$$
$$35 = 35$$

Both sides of the equation are equal, so $x = 45$.

Example 2 In the formula $a = bc$, find the value of c if $a = 20$ and $b = 2$.

Step 1 Rearrange the formula so the unknown value is by itself on one side of the equation by dividing both sides by b.
$$a = bc$$
$$\frac{a}{b} = \frac{bc}{b}$$
$$\frac{a}{b} = c$$

Step 2 Replace the variables a and b with the values that are given.
$$\frac{a}{b} = c$$
$$\frac{20}{2} = c$$
$$10 = c$$

Step 3 Check the solution.
$$a = bc$$
$$20 = 2 \times 10$$
$$20 = 20$$

Both sides of the equation are equal, so $c = 10$ is the solution when $a = 20$ and $b = 2$.

Practice Problem In the formula $h = gd$, find the value of d if $g = 12.3$ and $h = 17.4$.

Use Statistics

The branch of mathematics that deals with collecting, analyzing, and presenting data is statistics. In statistics, there are three common ways to summarize data with a single number—the mean, the median, and the mode.

The **mean** of a set of data is the arithmetic average. It is found by adding the numbers in the data set and dividing by the number of items in the set.

The **median** is the middle number in a set of data when the data are arranged in numerical order. If there were an even number of data points, the median would be the mean of the two middle numbers.

The **mode** of a set of data is the number or item that appears most often.

Another number that often is used to describe a set of data is the range. The **range** is the difference between the largest number and the smallest number in a set of data.

A **frequency table** shows how many times each piece of data occurs, usually in a survey. **Table 2** below shows the results of a student survey on favorite color.

Table 2 Student Color Choice		
Color	**Tally**	**Frequency**
red	IIII	4
blue	HHH	5
black	II	2
green	III	3
purple	HHH II	7
yellow	HHH I	6

Based on the frequency table data, which color is the favorite?

Example The speeds (in m/s) for a race car during five different time trials are 39, 37, 44, 36, and 44.

To find the mean:

Step 1 Find the sum of the numbers.

$$39 + 37 + 44 + 36 + 44 = 200$$

Step 2 Divide the sum by the number of items, which is 5.

$$200 \div 5 = 40$$

The mean is 40 m/s.

To find the median:

Step 1 Arrange the measures from least to greatest.

36, 37, 39, 44, 44

Step 2 Determine the middle measure.

36, 37, <u>39</u>, 44, 44

The median is 39 m/s.

To find the mode:

Step 1 Group the numbers that are the same together.

44, 44, 36, 37, 39

Step 2 Determine the number that occurs most in the set.

<u>44, 44</u>, 36, 37, 39

The mode is 44 m/s.

To find the range:

Step 1 Arrange the measures from largest to smallest.

44, 44, 39, 37, 36

Step 2 Determine the largest and smallest measures in the set.

<u>44</u>, 44, 39, 37, <u>36</u>

Step 3 Find the difference between the largest and smallest measures.

$$44 - 36 = 8$$

The range is 8 m/s.

Practice Problem Find the mean, median, mode, and range for the data set 8, 4, 12, 8, 11, 14, 16.

Use Geometry

The branch of mathematics that deals with the measurement, properties, and relationships of points, lines, angles, surfaces, and solids is called geometry.

Perimeter The **perimeter** (P) is the distance around a geometric figure. To find the perimeter of a rectangle, add the length and width and multiply that sum by two, or $2(l + w)$. To find perimeters of irregular figures, add the length of the sides.

Example 1 Find the perimeter of a rectangle that is 3 m long and 5 m wide.

Step 1 You know that the perimeter is 2 times the sum of the width and length.
$$P = 2(3\text{ m} + 5\text{ m})$$

Step 2 Find the sum of the width and length.
$$P = 2(8\text{ m})$$

Step 3 Multiply by 2.
$$P = 16\text{ m}$$

The perimeter is 16 m.

Example 2 Find the perimeter of a shape with sides measuring 2 cm, 5 cm, 6 cm, 3 cm.

Step 1 You know that the perimeter is the sum of all the sides.
$$P = 2 + 5 + 6 + 3$$

Step 2 Find the sum of the sides.
$$P = 2 + 5 + 6 + 3$$
$$P = 16$$

The perimeter is 16 cm.

Practice Problem Find the perimeter of a rectangle with a length of 18 m and a width of 7 m.

Practice Problem Find the perimeter of a triangle measuring 1.6 cm by 2.4 cm by 2.4 cm.

Area of a Rectangle The **area** (A) is the number of square units needed to cover a surface. To find the area of a rectangle, multiply the length times the width, or $l \times w$. When finding area, the units also are multiplied. Area is given in square units.

Example Find the area of a rectangle with a length of 1 cm and a width of 10 cm.

Step 1 You know that the area is the length multiplied by the width.
$$A = (1\text{ cm} \times 10\text{ cm})$$

Step 2 Multiply the length by the width. Also multiply the units.
$$A = 10\text{ cm}^2$$

The area is 10 cm^2.

Practice Problem Find the area of a square whose sides measure 4 m.

Area of a Triangle To find the area of a triangle, use the formula:

$$A = \frac{1}{2}(\text{base} \times \text{height})$$

The base of a triangle can be any of its sides. The height is the perpendicular distance from a base to the opposite endpoint, or vertex.

Example Find the area of a triangle with a base of 18 m and a height of 7 m.

Step 1 You know that the area is $\frac{1}{2}$ the base times the height.
$$A = \frac{1}{2}(18\text{ m} \times 7\text{ m})$$

Step 2 Multiply $\frac{1}{2}$ by the product of 18×7. Multiply the units.
$$A = \frac{1}{2}(126\text{ m}^2)$$
$$A = 63\text{ m}^2$$

The area is 63 m^2.

Practice Problem Find the area of a triangle with a base of 27 cm and a height of 17 cm.

Circumference of a Circle The **diameter** (d) of a circle is the distance across the circle through its center, and the **radius** (r) is the distance from the center to any point on the circle. The radius is half of the diameter. The distance around the circle is called the **circumference** (C). The formula for finding the circumference is:

$$C = 2\pi r \ \ or \ \ C = \pi d$$

The circumference divided by the diameter is always equal to 3.1415926... This nonterminating and nonrepeating number is represented by the Greek letter π (pi). An approximation often used for π is 3.14.

Example 1 Find the circumference of a circle with a radius of 3 m.

Step 1 You know the formula for the circumference is 2 times the radius times π.
$$C = 2\pi(3)$$

Step 2 Multiply 2 times the radius.
$$C = 6\pi$$

Step 3 Multiply by π.
$$C = 19 \text{ m}$$

The circumference is 19 m.

Example 2 Find the circumference of a circle with a diameter of 24.0 cm.

Step 1 You know the formula for the circumference is the diameter times π.
$$C = \pi(24.0)$$

Step 2 Multiply the diameter by π.
$$C = 75.4 \text{ cm}$$

The circumference is 75.4 cm.

Practice Problem Find the circumference of a circle with a radius of 19 cm.

Area of a Circle The formula for the area of a circle is:
$$A = \pi r^2$$

Example 1 Find the area of a circle with a radius of 4.0 cm.

Step 1 $A = \pi(4.0)^2$

Step 2 Find the square of the radius.
$$A = 16\pi$$

Step 3 Multiply the square of the radius by π.
$$A = 50 \text{ cm}^2$$

The area of the circle is 50 cm^2.

Example 2 Find the area of a circle with a radius of 225 m.

Step 1 $A = \pi(225)^2$

Step 2 Find the square of the radius.
$$A = 50625\pi$$

Step 3 Multiply the square of the radius by π.
$$A = 158962.5$$

The area of the circle is 158,962 m^2.

Example 3 Find the area of a circle whose diameter is 20.0 mm.

Step 1 You know the formula for the area of a circle is the square of the radius times π, and that the radius is half of the diameter.
$$A = \pi\left(\frac{20.0}{2}\right)^2$$

Step 2 Find the radius.
$$A = \pi(10.0)^2$$

Step 3 Find the square of the radius.
$$A = 100\pi$$

Step 4 Multiply the square of the radius by π.
$$A = 314 \text{ mm}^2$$

The area is 314 mm^2.

Practice Problem Find the area of a circle with a radius of 16 m.

Volume The measure of space occupied by a solid is the **volume** (*V*). To find the volume of a rectangular solid multiply the length times width times height, or $V = l \times w \times h$. It is measured in cubic units, such as cubic centimeters (cm^3).

Example Find the volume of a rectangular solid with a length of 2.0 m, a width of 4.0 m, and a height of 3.0 m.

Step 1 You know the formula for volume is the length times the width times the height.
$$V = 2.0 \text{ m} \times 4.0 \text{ m} \times 3.0 \text{ m}$$

Step 2 Multiply the length times the width times the height.
$$V = 24 \text{ m}^3$$

The volume is 24 m^3.

Practice Problem Find the volume of a rectangular solid that is 8 m long, 4 m wide, and 4 m high.

To find the volume of other solids, multiply the area of the base times the height.

Example 1 Find the volume of a solid that has a triangular base with a length of 8.0 m and a height of 7.0 m. The height of the entire solid is 15.0 m.

Step 1 You know that the base is a triangle, and the area of a triangle is $\frac{1}{2}$ the base times the height, and the volume is the area of the base times the height.
$$V = \left[\frac{1}{2} (b \times h) \right] \times 15$$

Step 2 Find the area of the base.
$$V = \left[\frac{1}{2} (8 \times 7) \right] \times 15$$
$$V = \left(\frac{1}{2} \times 56 \right) \times 15$$

Step 3 Multiply the area of the base by the height of the solid.
$$V = 28 \times 15$$
$$V = 420 \text{ m}^3$$

The volume is 420 m^3.

Example 2 Find the volume of a cylinder that has a base with a radius of 12.0 cm, and a height of 21.0 cm.

Step 1 You know that the base is a circle, and the area of a circle is the square of the radius times π, and the volume is the area of the base times the height.
$$V = (\pi r^2) \times 21$$
$$V = (\pi 12^2) \times 21$$

Step 2 Find the area of the base.
$$V = 144\pi \times 21$$
$$V = 452 \times 21$$

Step 3 Multiply the area of the base by the height of the solid.
$$V = 9490 \text{ cm}^3$$

The volume is 9490 cm^3.

Example 3 Find the volume of a cylinder that has a diameter of 15 mm and a height of 4.8 mm.

Step 1 You know that the base is a circle with an area equal to the square of the radius times π. The radius is one-half the diameter. The volume is the area of the base times the height.
$$V = (\pi r^2) \times 4.8$$
$$V = \left[\pi \left(\frac{1}{2} \times 15 \right)^2 \right] \times 4.8$$
$$V = (\pi 7.5^2) \times 4.8$$

Step 2 Find the area of the base.
$$V = 56.25\pi \times 4.8$$
$$V = 176.63 \times 4.8$$

Step 3 Multiply the area of the base by the height of the solid.
$$V = 847.8$$

The volume is 847.8 mm^3.

Practice Problem Find the volume of a cylinder with a diameter of 7 cm in the base and a height of 16 cm.

Science Applications

Measure in SI

The metric system of measurement was developed in 1795. A modern form of the metric system, called the International System (SI), was adopted in 1960 and provides the standard measurements that all scientists around the world can understand.

The SI system is convenient because unit sizes vary by powers of 10. Prefixes are used to name units. Look at **Table 3** for some common SI prefixes and their meanings.

Table 3 Common SI Prefixes			
Prefix	**Symbol**	**Meaning**	
kilo-	k	1,000	thousand
hecto-	h	100	hundred
deka-	da	10	ten
deci-	d	0.1	tenth
centi-	c	0.01	hundredth
milli-	m	0.001	thousandth

Example How many grams equal one kilogram?

Step 1 Find the prefix *kilo* in **Table 3.**

Step 2 Using **Table 3,** determine the meaning of *kilo.* According to the table, it means 1,000. When the prefix *kilo* is added to a unit, it means that there are 1,000 of the units in a "*kilo*unit."

Step 3 Apply the prefix to the units in the question. The units in the question are grams. There are 1,000 grams in a kilogram.

Practice Problem Is a milligram larger or smaller than a gram? How many of the smaller units equal one larger unit? What fraction of the larger unit does one smaller unit represent?

Dimensional Analysis

Convert SI Units In science, quantities such as length, mass, and time sometimes are measured using different units. A process called dimensional analysis can be used to change one unit of measure to another. This process involves multiplying your starting quantity and units by one or more conversion factors. A conversion factor is a ratio equal to one and can be made from any two equal quantities with different units. If 1,000 mL equal 1 L then two ratios can be made.

$$\frac{1,000 \text{ mL}}{1 \text{ L}} = \frac{1 \text{ L}}{1,000 \text{ mL}} = 1$$

One can covert between units in the SI system by using the equivalents in **Table 3** to make conversion factors.

Example 1 How many cm are in 4 m?

Step 1 Write conversion factors for the units given. From **Table 3,** you know that 100 cm = 1 m. The conversion factors are

$$\frac{100 \text{ cm}}{1 \text{ m}} \quad and \quad \frac{1 \text{ m}}{100 \text{ cm}}$$

Step 2 Decide which conversion factor to use. Select the factor that has the units you are converting from (m) in the denominator and the units you are converting to (cm) in the numerator.

$$\frac{100 \text{ cm}}{1 \text{ m}}$$

Step 3 Multiply the starting quantity and units by the conversion factor. Cancel the starting units with the units in the denominator. There are 400 cm in 4 m.

$$4 \text{ m} \times \frac{100 \text{ cm}}{1 \text{ m}} = 400 \text{ cm}$$

Practice Problem How many milligrams are in one kilogram? (Hint: You will need to use two conversion factors from **Table 3.**)

Table 4 Unit System Equivalents

Type of Measurement	Equivalent
Length	1 in = 2.54 cm
	1 yd = 0.91 m
	1 mi = 1.61 km
Mass and Weight*	1 oz = 28.35 g
	1 lb = 0.45 kg
	1 ton (short) = 0.91 tonnes (metric tons)
	1 lb = 4.45 N
Volume	$1 \text{ in}^3 = 16.39 \text{ cm}^3$
	1 qt = 0.95 L
	1 gal = 3.78 L
Area	$1 \text{ in}^2 = 6.45 \text{ cm}^2$
	$1 \text{ yd}^2 = 0.83 \text{ m}^2$
	$1 \text{ mi}^2 = 2.59 \text{ km}^2$
	1 acre = 0.40 hectares
Temperature	$^\circ C = \dfrac{(^\circ F - 32)}{1.8}$
	$K = {}^\circ C + 273$

*Weight is measured in standard Earth gravity.

Convert Between Unit Systems Table 4 gives a list of equivalents that can be used to convert between English and SI units.

Example If a meterstick has a length of 100 cm, how long is the meterstick in inches?

Step 1 Write the conversion factors for the units given. From **Table 4,** 1 in = 2.54 cm.

$$\frac{1 \text{ in}}{2.54 \text{ cm}} \quad and \quad \frac{2.54 \text{ cm}}{1 \text{ in}}$$

Step 2 Determine which conversion factor to use. You are converting from cm to in. Use the conversion factor with cm on the bottom.

$$\frac{1 \text{ in}}{2.54 \text{ cm}}$$

Step 3 Multiply the starting quantity and units by the conversion factor. Cancel the starting units with the units in the denominator. Round your answer based on the number of significant figures in the conversion factor.

$$100 \cancel{\text{ cm}} \times \frac{1 \text{ in}}{2.54 \cancel{\text{ cm}}} = 39.37 \text{ in}$$

The meterstick is 39.4 in long.

Practice Problem A book has a mass of 5 lbs. What is the mass of the book in kg?

Practice Problem Use the equivalent for in and cm (1 in = 2.54 cm) to show how $1 \text{ in}^3 = 16.39 \text{ cm}^3$.

Precision and Significant Digits

When you make a measurement, the value you record depends on the precision of the measuring instrument. This precision is represented by the number of significant digits recorded in the measurement. When counting the number of significant digits, all digits are counted except zeros at the end of a number with no decimal point such as 2,050, and zeros at the beginning of a decimal such as 0.03020. When adding or subtracting numbers with different precision, round the answer to the smallest number of decimal places of any number in the sum or difference. When multiplying or dividing, the answer is rounded to the smallest number of significant digits of any number being multiplied or divided.

Example The lengths 5.28 and 5.2 are measured in meters. Find the sum of these lengths and record your answer using the correct number of significant digits.

Step 1 Find the sum.

5.28 m	2 digits after the decimal
+ 5.2 m	1 digit after the decimal
10.48 m	

Step 2 Round to one digit after the decimal because the least number of digits after the decimal of the numbers being added is 1.

The sum is 10.5 m.

Practice Problem How many significant digits are in the measurement 7,071,301 m? How many significant digits are in the measurement 0.003010 g?

Practice Problem Multiply 5.28 and 5.2 using the rule for multiplying and dividing. Record the answer using the correct number of significant digits.

Scientific Notation

Many times numbers used in science are very small or very large. Because these numbers are difficult to work with scientists use scientific notation. To write numbers in scientific notation, move the decimal point until only one non-zero digit remains on the left. Then count the number of places you moved the decimal point and use that number as a power of ten. For example, the average distance from the Sun to Mars is 227,800,000,000 m. In scientific notation, this distance is 2.278×10^{11} m. Because you moved the decimal point to the left, the number is a positive power of ten.

The mass of an electron is about 0.000 000 000 000 000 000 000 000 000 000 911 kg. Expressed in scientific notation, this mass is 9.11×10^{-31} kg. Because the decimal point was moved to the right, the number is a negative power of ten.

Example Earth is 149,600,000 km from the Sun. Express this in scientific notation.

Step 1 Move the decimal point until one non-zero digit remains on the left.
1.496 000 00

Step 2 Count the number of decimal places you have moved. In this case, eight.

Step 3 Show that number as a power of ten, 10^8.

The Earth is 1.496×10^8 km from the Sun.

Practice Problem How many significant digits are in 149,600,000 km? How many significant digits are in 1.496×10^8 km?

Practice Problem Parts used in a high performance car must be measured to 7×10^{-6} m. Express this number as a decimal.

Practice Problem A CD is spinning at 539 revolutions per minute. Express this number in scientific notation.

Make and Use Graphs

Data in tables can be displayed in a graph—a visual representation of data. Common graph types include line graphs, bar graphs, and circle graphs.

Line Graph A line graph shows a relationship between two variables that change continuously. The independent variable is changed and is plotted on the *x*-axis. The dependent variable is observed, and is plotted on the *y*-axis.

Example Draw a line graph of the data below from a cyclist in a long-distance race.

Table 5 Bicycle Race Data	
Time (h)	**Distance (km)**
0	0
1	8
2	16
3	24
4	32
5	40

Step 1 Determine the *x*-axis and *y*-axis variables. Time varies independently of distance and is plotted on the *x*-axis. Distance is dependent on time and is plotted on the *y*-axis.

Step 2 Determine the scale of each axis. The *x*-axis data ranges from 0 to 5. The *y*-axis data ranges from 0 to 40.

Step 3 Using graph paper, draw and label the axes. Include units in the labels.

Step 4 Draw a point at the intersection of the time value on the *x*-axis and corresponding distance value on the *y*-axis. Connect the points and label the graph with a title, as shown in **Figure 20.**

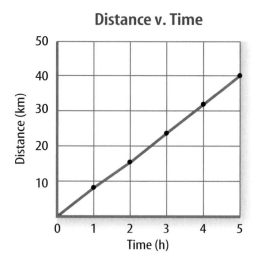

Distance v. Time

Figure 20 This line graph shows the relationship between distance and time during a bicycle ride.

Practice Problem A puppy's shoulder height is measured during the first year of her life. The following measurements were collected: (3 mo, 52 cm), (6 mo, 72 cm), (9 mo, 83 cm), (12 mo, 86 cm). Graph this data.

Find a Slope The slope of a straight line is the ratio of the vertical change, rise, to the horizontal change, run.

$$\text{Slope} = \frac{\text{vertical change (rise)}}{\text{horizontal change (run)}} = \frac{\text{change in } y}{\text{change in } x}$$

Example Find the slope of the graph in **Figure 20.**

Step 1 You know that the slope is the change in *y* divided by the change in *x*.

$$\text{Slope} = \frac{\text{change in } y}{\text{change in } x}$$

Step 2 Determine the data points you will be using. For a straight line, choose the two sets of points that are the farthest apart.

$$\text{Slope} = \frac{(40-0) \text{ km}}{(5-0) \text{ hr}}$$

Step 3 Find the change in *y* and *x*.

$$\text{Slope} = \frac{40 \text{ km}}{5 \text{ h}}$$

Step 4 Divide the change in *y* by the change in *x*.

$$\text{Slope} = \frac{8 \text{ km}}{\text{h}}$$

The slope of the graph is 8 km/h.

Bar Graph To compare data that does not change continuously you might choose a bar graph. A bar graph uses bars to show the relationships between variables. The *x*-axis variable is divided into parts. The parts can be numbers such as years, or a category such as a type of animal. The *y*-axis is a number and increases continuously along the axis.

Example A recycling center collects 4.0 kg of aluminum on Monday, 1.0 kg on Wednesday, and 2.0 kg on Friday. Create a bar graph of this data.

Step 1 Select the *x*-axis and *y*-axis variables. The measured numbers (the masses of aluminum) should be placed on the *y*-axis. The variable divided into parts (collection days) is placed on the *x*-axis.

Step 2 Create a graph grid like you would for a line graph. Include labels and units.

Step 3 For each measured number, draw a vertical bar above the *x*-axis value up to the *y*-axis value. For the first data point, draw a vertical bar above Monday up to 4.0 kg.

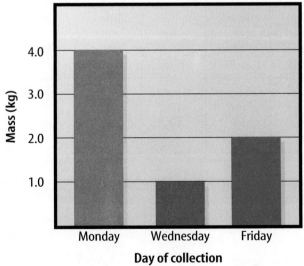

Aluminum Collected During Week

Practice Problem Draw a bar graph of the gases in air: 78% nitrogen, 21% oxygen, 1% other gases.

Circle Graph To display data as parts of a whole, you might use a circle graph. A circle graph is a circle divided into sections that represent the relative size of each piece of data. The entire circle represents 100%, half represents 50%, and so on.

Example Air is made up of 78% nitrogen, 21% oxygen, and 1% other gases. Display the composition of air in a circle graph.

Step 1 Multiply each percent by 360° and divide by 100 to find the angle of each section in the circle.

$$78\% \times \frac{360°}{100} = 280.8°$$

$$21\% \times \frac{360°}{100} = 75.6°$$

$$1\% \times \frac{360°}{100} = 3.6°$$

Step 2 Use a compass to draw a circle and to mark the center of the circle. Draw a straight line from the center to the edge of the circle.

Step 3 Use a protractor and the angles you calculated to divide the circle into parts. Place the center of the protractor over the center of the circle and line the base of the protractor over the straight line.

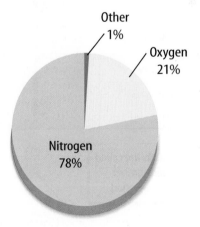

Practice Problem Draw a circle graph to represent the amount of aluminum collected during the week shown in the bar graph to the left.

PERIODIC TABLE OF THE ELEMENTS

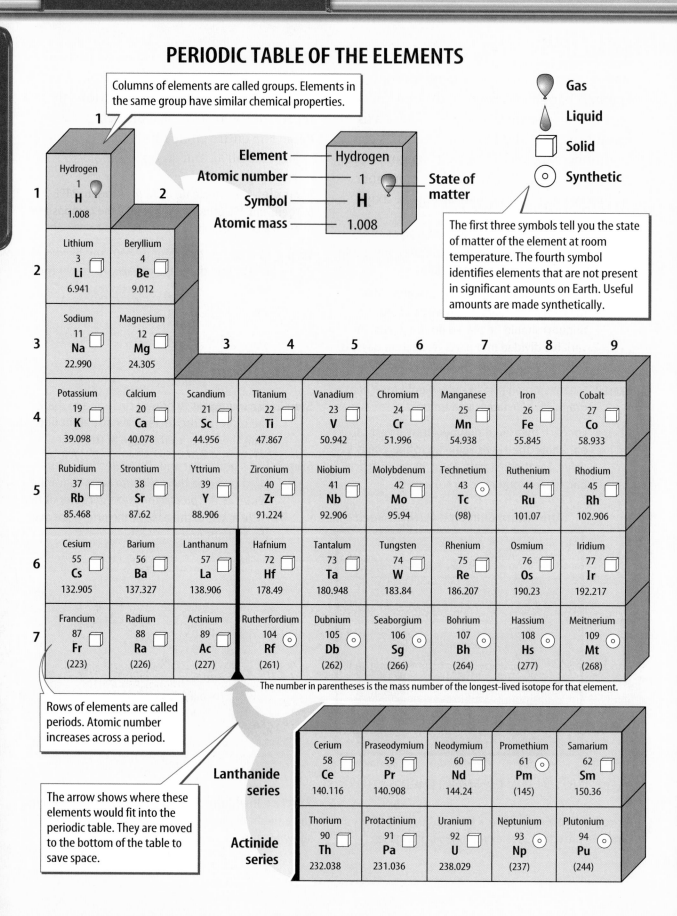

Metal

Metalloid

Nonmetal

The color of an element's block tells you if the element is a metal, nonmetal, or metalloid.

Science Online

Visit booke.msscience.com for updates to the periodic table.

18

| Helium | 2 | **He** | 4.003 |

13 | **14** | **15** | **16** | **17**

Boron	5	**B**	10.811
Carbon	6	**C**	12.011
Nitrogen	7	**N**	14.007
Oxygen	8	**O**	15.999
Fluorine	9	**F**	18.998
Neon	10	**Ne**	20.180

Aluminum	13	**Al**	26.982
Silicon	14	**Si**	28.086
Phosphorus	15	**P**	30.974
Sulfur	16	**S**	32.065
Chlorine	17	**Cl**	35.453
Argon	18	**Ar**	39.948

10 | **11** | **12**

Nickel	28	**Ni**	58.693
Copper	29	**Cu**	63.546
Zinc	30	**Zn**	65.409
Gallium	31	**Ga**	69.723
Germanium	32	**Ge**	72.64
Arsenic	33	**As**	74.922
Selenium	34	**Se**	78.96
Bromine	35	**Br**	79.904
Krypton	36	**Kr**	83.798

Palladium	46	**Pd**	106.42
Silver	47	**Ag**	107.868
Cadmium	48	**Cd**	112.411
Indium	49	**In**	114.818
Tin	50	**Sn**	118.710
Antimony	51	**Sb**	121.760
Tellurium	52	**Te**	127.60
Iodine	53	**I**	126.904
Xenon	54	**Xe**	131.293

Platinum	78	**Pt**	195.078
Gold	79	**Au**	196.967
Mercury	80	**Hg**	200.59
Thallium	81	**Tl**	204.383
Lead	82	**Pb**	207.2
Bismuth	83	**Bi**	208.980
Polonium	84	**Po**	(209)
Astatine	85	**At**	(210)
Radon	86	**Rn**	(222)

Darmstadtium	110	**Ds**	(281)
Unununium	* 111	**Uuu**	(272)
Ununbium	* 112	**Uub**	(285)
Ununquadium	* 114	**Uuq**	(289)
** 116
** 118

* The names and symbols for elements 111–114 are temporary. Final names will be selected when the elements' discoveries are verified.

** Elements 116 and 118 were thought to have been created. The claim was retracted because the experimental results could not be repeated.

Europium	63	**Eu**	151.964
Gadolinium	64	**Gd**	157.25
Terbium	65	**Tb**	158.925
Dysprosium	66	**Dy**	162.500
Holmium	67	**Ho**	164.930
Erbium	68	**Er**	167.259
Thulium	69	**Tm**	168.934
Ytterbium	70	**Yb**	173.04
Lutetium	71	**Lu**	174.967

Americium	95	**Am**	(243)
Curium	96	**Cm**	(247)
Berkelium	97	**Bk**	(247)
Californium	98	**Cf**	(251)
Einsteinium	99	**Es**	(252)
Fermium	100	**Fm**	(257)
Mendelevium	101	**Md**	(258)
Nobelium	102	**No**	(259)
Lawrencium	103	**Lr**	(262)

Use and Care of a Microscope

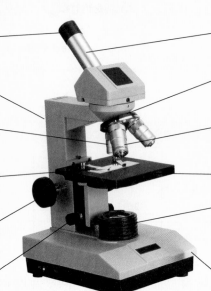

Eyepiece Contains magnifying lenses you look through.

Arm Supports the body tube.

Low-power objective Contains the lens with the lowest power magnification.

Stage clips Hold the microscope slide in place.

Coarse adjustment Focuses the image under low power.

Fine adjustment Sharpens the image under high magnification.

Body tube Connects the eyepiece to the revolving nosepiece.

Revolving nosepiece Holds and turns the objectives into viewing position.

High-power objective Contains the lens with the highest magnification.

Stage Supports the microscope slide.

Light source Provides light that passes upward through the diaphragm, the specimen, and the lenses.

Base Provides support for the microscope.

Caring for a Microscope

1. Always carry the microscope holding the arm with one hand and supporting the base with the other hand.

2. Don't touch the lenses with your fingers.

3. The coarse adjustment knob is used only when looking through the lowest-power objective lens. The fine adjustment knob is used when the high-power objective is in place.

4. Cover the microscope when you store it.

Using a Microscope

1. Place the microscope on a flat surface that is clear of objects. The arm should be toward you.

2. Look through the eyepiece. Adjust the diaphragm so light comes through the opening in the stage.

3. Place a slide on the stage so the specimen is in the field of view. Hold it firmly in place by using the stage clips.

4. Always focus with the coarse adjustment and the low-power objective lens first. After the object is in focus on low power, turn the nosepiece until the high-power objective is in place. Use ONLY the fine adjustment to focus with the high-power objective lens.

Making a Wet-Mount Slide

1. Carefully place the item you want to look at in the center of a clean, glass slide. Make sure the sample is thin enough for light to pass through.

2. Use a dropper to place one or two drops of water on the sample.

3. Hold a clean coverslip by the edges and place it at one edge of the water. Slowly lower the coverslip onto the water until it lies flat.

4. If you have too much water or a lot of air bubbles, touch the edge of a paper towel to the edge of the coverslip to draw off extra water and draw out unwanted air.

Diversity of Life: Classification of Living Organisms

A six-kingdom system of classification of organisms is used today. Two kingdoms—Kingdom Archaebacteria and Kingdom Eubacteria—contain organisms that do not have a nucleus and that lack membrane-bound structures in the cytoplasm of their cells. The members of the other four kingdoms have a cell or cells that contain a nucleus and structures in the cytoplasm, some of which are surrounded by membranes. These kingdoms are Kingdom Protista, Kingdom Fungi, Kingdom Plantae, and Kingdom Animalia.

Kingdom Archaebacteria

one-celled; some absorb food from their surroundings; some are photosynthetic; some are chemosynthetic; many are found in extremely harsh environments including salt ponds, hot springs, swamps, and deep-sea hydrothermal vents

Kingdom Eubacteria

one-celled; most absorb food from their surroundings; some are photosynthetic; some are chemosynthetic; many are parasites; many are round, spiral, or rod-shaped; some form colonies

Kingdom Protista

Phylum Euglenophyta one-celled; photosynthetic or take in food; most have one flagellum; euglenoids

Phylum Bacillariophyta one-celled; photosynthetic; have unique double shells made of silica; diatoms

Phylum Dinoflagellata one-celled; photosynthetic; contain red pigments; have two flagella; dinoflagellates

Phylum Chlorophyta one-celled, many-celled, or colonies; photosynthetic; contain chlorophyll; live on land, in freshwater, or salt water; green algae

Phylum Rhodophyta most are many-celled; photosynthetic; contain red pigments; most live in deep, saltwater environments; red algae

Phylum Phaeophyta most are many-celled; photosynthetic; contain brown pigments; most live in saltwater environments; brown algae

Phylum Rhizopoda one-celled; take in food; are free-living or parasitic; move by means of pseudopods; amoebas

Kingdom Eubacteria
Bacillus anthracis

Phylum Chlorophyta
Desmids

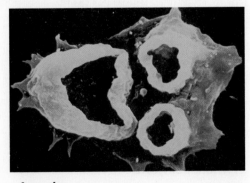

Amoeba

Phylum Zoomastigina one-celled; take in food; free-living or parasitic; have one or more flagella; zoomastigotes

Phylum Ciliophora one-celled; take in food; have large numbers of cilia; ciliates

Phylum Sporozoa one-celled; take in food; have no means of movement; are parasites in animals; sporozoans

Phyla Myxomycota and Acrasiomycota one- or many-celled; absorb food; change form during life cycle; cellular and plasmodial slime molds

Phylum Oomycota many-celled; are either parasites or decomposers; live in freshwater or salt water; water molds, rusts and downy mildews

Kingdom Fungi

Phylum Zygomycota many-celled; absorb food; spores are produced in sporangia; zygote fungi; bread mold

Phylum Ascomycota one- and many-celled; absorb food; spores produced in asci; sac fungi; yeast

Phylum Basidiomycota many-celled; absorb food; spores produced in basidia; club fungi; mushrooms

Phylum Deuteromycota members with unknown reproductive structures; imperfect fungi; *Penicillium*

Phylum Mycophycota organisms formed by symbiotic relationship between an ascomycote or a basidiomycote and green alga or cyanobacterium; lichens

Phylum Myxomycota
Slime mold

Phylum Oomycota
Phytophthora infestans

Lichens

Kingdom Plantae

Divisions Bryophyta (mosses), **Anthocerophyta** (hornworts), **Hepaticophyta** (liverworts), **Psilophyta** (whisk ferns) many-celled nonvascular plants; reproduce by spores produced in capsules; green; grow in moist, land environments

Division Lycophyta many-celled vascular plants; spores are produced in conelike structures; live on land; are photosynthetic; club mosses

Division Arthrophyta vascular plants; ribbed and jointed stems; scalelike leaves; spores produced in conelike structures; horsetails

Division Pterophyta vascular plants; leaves called fronds; spores produced in clusters of sporangia called sori; live on land or in water; ferns

Division Ginkgophyta deciduous trees; only one living species; have fan-shaped leaves with branching veins and fleshy cones with seeds; ginkgoes

Division Cycadophyta palmlike plants; have large, featherlike leaves; produces seeds in cones; cycads

Division Coniferophyta deciduous or evergreen; trees or shrubs; have needlelike or scalelike leaves; seeds produced in cones; conifers

Division Anthophyta
Tomato plant

Division Gnetophyta shrubs or woody vines; seeds are produced in cones; division contains only three genera; gnetum

Division Anthophyta dominant group of plants; flowering plants; have fruits with seeds

Kingdom Animalia

Phylum Porifera aquatic organisms that lack true tissues and organs; are asymmetrical and sessile; sponges

Phylum Cnidaria radially symmetrical organisms; have a digestive cavity with one opening; most have tentacles armed with stinging cells; live in aquatic environments singly or in colonies; includes jellyfish, corals, hydra, and sea anemones

Phylum Platyhelminthes bilaterally symmetrical worms; have flattened bodies; digestive system has one opening; parasitic and free-living species; flatworms

Division Bryophyta
Liverwort

Phylum Platyhelminthes
Flatworm

Phylum Chordata

Phylum Nematoda round, bilaterally symmetrical body; have digestive system with two openings; free-living forms and parasitic forms; roundworms

Phylum Mollusca soft-bodied animals, many with a hard shell and soft foot or footlike appendage; a mantle covers the soft body; aquatic and terrestrial species; includes clams, snails, squid, and octopuses

Phylum Annelida bilaterally symmetrical worms; have round, segmented bodies; terrestrial and aquatic species; includes earthworms, leeches, and marine polychaetes

Phylum Arthropoda largest animal group; have hard exoskeletons, segmented bodies, and pairs of jointed appendages; land and aquatic species; includes insects, crustaceans, and spiders

Phylum Echinodermata marine organisms; have spiny or leathery skin and a water-vascular system with tube feet; are radially symmetrical; includes sea stars, sand dollars, and sea urchins

Phylum Chordata organisms with internal skeletons and specialized body systems; most have paired appendages; all at some time have a notochord, nerve cord, gill slits, and a post-anal tail; include fish, amphibians, reptiles, birds, and mammals

Cómo usar el glosario en español:
1. Busca el término en inglés que desees encontrar.
2. El término en español, junto con la definición, se encuentran en la columna de la derecha.

Pronunciation Key

Use the following key to help you sound out words in the glossary.

a..............	back (BAK)	ew	food (FEWD)
ay..............	day (DAY)	yoo	pure (PYOOR)
ah..............	father (FAH thur)	yew	few (FYEW)
ow	flower (FLOW ur)	uh	comma (CAH muh)
ar	car (CAR)	u (+ con)......	rub (RUB)
e..............	less (LES)	sh..............	shelf (SHELF)
ee..............	leaf (LEEF)	ch..............	nature (NAY chur)
ih	trip (TRIHP)	g..............	gift (GIHFT)
i (i + con + e) ..	idea (i DEE uh)	j	gem (JEM)
oh	go (GOH)	ing..............	sing (SING)
aw	soft (SAWFT)	zh..............	vision (VIH zhun)
or	orbit (OR buht)	k..............	cake (KAYK)
oy..............	coin (COYN)	s	seed, cent (SEED, SENT)
oo	foot (FOOT)	z..............	zone, raise (ZOHN, RAYZ)

English — A — **Español**

abiotic: nonliving, physical features of the environment, including air, water, sunlight, soil, temperature, and climate. (p. 36)

acid precipitation: precipitation with a pH below 5.6—which occurs when air pollutants from the burning of fossil fuels react with water in the atmosphere to form strong acids—that can pollute water, kill fish and plants, and damage soils. (p. 103)

acid rain: forms when sulfur dioxide and nitrogen oxide from industries and car exhausts combine with water vapor in the air; can wash nutrients from soil and damage trees and aquatic life. (p. 135)

atmosphere: air surrounding Earth; is made up of gases, including 78 percent nitrogen, 21 percent oxygen, and 0.03 percent carbon dioxide. (p. 37)

abiótico: características inertes y físicas del medio ambiente, incluyendo el aire, el agua, la luz solar, el suelo, la temperatura y el clima. (p. 36)

lluvia ácida: precipitación con un pH menor de 5.6—lo cual ocurre cuando los contaminantes del aire provenientes de la quema de combustibles fósiles reaccionan con el agua en la atmósfera para formar ácidos fuertes—que puede contaminar el agua, matar peces y plantas, y dañar los suelos. (p. 103)

lluvia ácida: se forma cuando el dióxido de azufre y el óxido de nitrógeno derivados de la industria y de los escapes de los automóviles se combinan con vapor de agua en el aire; puede arrastrar nutrientes del suelo y causar daño a los árboles y a la vida acuática. (p. 135)

atmósfera: aire que rodea a la Tierra; está compuesta de gases, incluyendo 78% de nitrógeno, 21% de oxígeno y 0.03% de dióxido de carbono. (p. 37)

— B —

biodiversity: variety of life in an ecosystem, most commonly measured by the number of species that live in a given area. (p. 126)

biodiversidad: variedad de vida en un ecosistema, comúnmente cuantificada mediante el número de especies que viven en un área determinada. (p. 126)

Glossary/Glosario

biomes (BI ohmz): large geographic areas with similar climates and ecosystems; includes tundra, taiga, desert, temperate deciduous forest, temperate rain forest, tropical rain forest, and grassland. (p. 68)

biosphere: part of Earth that supports life, including the top portion of Earth's crust, the atmosphere, and all the water on Earth's surface. (p. 8)

biotic (bi AH tihk): features of the environment that are alive or were once alive. (p. 36)

biomas: grandes áreas geográficas con climas y ecosistemas similares; incluyen la tundra, la taiga, el desierto, el bosque caducifolio templado, el bosque lluvioso templado, la selva húmeda tropical y los pastizales. (p. 68)

biosfera: capa de la Tierra que alberga la vida, incluyendo la porción superior de la corteza terrestre, la atmósfera y toda el agua de la superficie terrestre. (p. 8)

biótico: características del ambiente que tienen o alguna vez tuvieron vida. (p. 36)

C

captive population: population of organisms that is cared for by humans. (p. 142)

carbon cycle: model describing how carbon molecules move between the living and nonliving world. (p. 49)

carrying capacity: largest number of individuals of a particular species that an ecosystem can support over time. (p. 15)

chemosynthesis (kee moh SIN thuh sus): process in which producers make energy-rich nutrient molecules from chemicals. (p. 51)

climate: average weather conditions of an area over time, including wind, temperature, and rainfall or other types of precipitation such as snow or sleet. (p. 41)

climax community: stable, end stage of ecological succession in which balance is in the absence of disturbance. (p. 67)

commensalism: a type of symbiotic relationship in which one organism benefits and the other organism is not affected. (p. 22)

community: all the populations of different species that live in an ecosystem. (p. 10)

condensation: process that takes place when a gas changes to a liquid. (p. 45)

conservation biology: study of methods for protecting Earth's biodiversity; uses strategies such as reintroduction programs and habitat restoration and works to preserve threatened and endangered species. (p. 138)

consumer: organism that cannot create energy-rich molecules but obtains its food by eating other organisms. (p. 21)

coral reef: diverse ecosystem formed from the calcium carbonate shells secreted by corals. (p. 81)

población cautiva: población de organismos bajo el cuidado de los seres humanos. (p. 142)

ciclo del carbono: modelo que describe cómo se mueven las moléculas de carbono entre el mundo vivo y el mundo inerte. (p. 49)

capacidad de carga: el mayor número de individuos de una especie en particular que un ecosistema puede albergar en un periodo de tiempo. (p. 15)

quimiosíntesis: proceso a través del cual los productores fabrican moléculas ricas en energía a partir de agentes químicos. (p. 51)

clima: condiciones meteorológicas promedio de un área durante un periodo de tiempo; incluye viento, temperatura y precipitación pluvial u otros tipos de precipitación como la nieve o el granizo. (p. 41)

clímax comunitario: etapa final estable de la sucesión ecológica en la cual se da un equilibrio en ausencia de alteraciones. (p. 67)

comensalismo: tipo de relación simbiótica en la que un organismo se beneficia sin afectar al otro. (p. 22)

comunidad: todas las poblaciones de diferentes especies que viven en un mismo ecosistema. (p. 10)

condensación: proceso que tiene lugar cuando un gas cambia a estado líquido. (p. 45)

biología de la conservación: estudio de los métodos para proteger la biodiversidad de la Tierra; utiliza estrategias tales como programas de reintroducción y restauración de hábitat y busca preservar especies amenazadas o en peligro de extinción. (p. 138)

consumidor: organismo que no puede fabricar moléculas ricas en energía por lo que debe obtener su alimento ingiriendo otros organismos. (p. 21)

arrecife de coral: ecosistema diverso conformado de caparazones de carbonato de calcio secretados por los corales. (p. 81)

D

desert: driest biome on Earth with less than 25 cm of rain each year; has dunes or thin soil with little organic matter, where plants and animals are adapted to survive extreme conditions. (p. 74)

desierto: el bioma más seco sobre la Tierra con menos de 25 centímetros cúbicos de lluvia al año; tiene dunas o un suelo delgado con muy poca materia orgánica y aquí las plantas y animales están adaptados para sobrevivir en condiciones extremosas. (p. 74)

E

ecology: study of the interactions that take place among organisms and their environment. (p. 9)

ecosystem: all the living organisms that live in an area and the nonliving features of their environment. (p. 9)

endangered species: species that is in danger of becoming extinct. (p. 131)

energy pyramid: model that shows the amount of energy available at each feeding level in an ecosystem. (p. 53)

erosion: movement of soil from one place to another. (p. 109)

estuary: extremely fertile area where a river meets an ocean; contains a mixture of freshwater and saltwater and serves as a nursery for many species of fish. (p. 82)

evaporation: process that takes place when a liquid changes to a gas. (p. 44)

extinct species: species that was once present on Earth but has died out. (p. 130)

ecología: estudio de las interacciones que se dan entre los organismos y su medio ambiente. (p. 9)

ecosistema: conjunto de organismos vivos que habitan en un área y las características de su medio ambiente. (p. 9)

especie en peligro de extinción: especies que se encuentran en peligro de quedar extintas. (p. 131)

pirámide de energía: modelo que muestra la cantidad de energía disponible en cada nivel alimenticio de un ecosistema. (p. 53)

erosión: movimiento del suelo de un lugar a otro. (p. 109)

estuario: área extremadamente fértil donde un río desemboca en el océano; contiene una mezcla de agua dulce y salada y sirve como vivero para muchas especies de peces. (p. 82)

evaporación: proceso que tiene lugar cuando un líquido cambia a estado gaseoso. (p. 44)

especies extintas: especies que alguna vez estuvieron presentes en la Tierra pero que han desaparecido. (p. 130)

F

food web: model that shows the complex feeding relationships among organisms in a community. (p. 52)

fossil fuels: nonrenewable energy resources—coal, oil, and natural gas—that formed in Earth's crust over hundreds of millions of years. (p. 96)

cadena alimenticia: modelo que muestra las complejas relaciones alimenticias entre los organismos de una comunidad. (p. 52)

combustibles fósiles: recursos energéticos no renovables—carbón, petróleo y gas natural—que se formaron en la corteza terrestre durante cientos de millones de años. (p. 96)

G

geothermal energy: heat energy within Earth's crust, available only where natural geysers or volcanoes are located. (p. 99)

energía geotérmica: energía calórica en el interior de la corteza terrestre disponible sólo donde existen géiseres o volcanes. (p. 99)

grasslands: temperate and tropical regions with 25 cm to 75 cm of precipitation each year that are dominated by climax communities of grasses; ideal for growing crops and raising cattle and sheep. (p. 75)

greenhouse effect: heat-trapping feature of the atmosphere that keeps Earth warm enough to support life. (p. 104)

pastizales: regiones tropicales y templadas con 25 a 75 centímetros cúbicos de lluvia al año; son dominadas por el clímax comunitario de los pastos e ideales para la cría de ganado y ovejas. (p. 75)

efecto de invernadero: característica de la atmósfera que le permite atrapar calor y mantener la Tierra lo suficientemente caliente para favorecer la vida. (p. 104)

H

habitat: place where an organism lives and that provides the types of food, shelter, moisture, and temperature needed for survival. (p. 11)

habitat restoration: process of bringing a damaged habitat back to a healthy condition. (p. 141)

hazardous wastes: waste materials, such as pesticides and leftover paints, that are harmful to human health or poisonous to living organisms. (p. 110)

hydroelectric power: electricity produced when the energy of falling water turns the blades of a generator turbine. (p. 97)

hábitat: lugar donde vive un organismo y que le proporciona los tipos de alimento, refugio, humedad y temperatura necesarios para su supervivencia. (p. 11)

restauración de hábitat: proceso de restaurar la condiciones favorables de un hábitat alterado. (p. 141)

desperdicios peligrosos: materiales de desecho como los pesticidas y residuos de pintura nocivos para la salud humana o dañinos para los organismos vivos. (p. 110)

energía hidroeléctrica: electricidad producida cuando la energía generada por la caída del agua hace girar las aspas de una turbina generadora. (p. 97)

I

intertidal zone: part of the shoreline that is under water at high tide and exposed to the air at low tide. (p. 82)

introduced species: species that moves into an ecosystem as a result of human actions. (p. 134)

zona litoral: parte de la línea costera que está bajo el agua durante la marea alta y expuesta al aire durante la marea baja. (p. 82)

especies introducidas: especies que ingresan en un ecosistema como resultado de las actividades humanas. (p. 134)

L

limiting factor: anything that can restrict the size of a population, including living and nonliving features of an ecosystem, such as predators or drought. (p. 14)

factor limitante: cualquier factor que pueda restringir el tamaño de una población, incluyendo las características biológicas y no biológicas de un ecosistema, tales como los depredadores o las sequías. (p. 14)

M

mutualism: a type of symbiotic relationship in which both organisms benefit. (p. 22)

mutualismo: tipo de relación simbiótica en la que ambos organismos se benefician. (p. 22)

N

native species: original organisms in an ecosystem. (p. 134)

natural resources: parts of Earth's environment that supply materials useful or necessary for the survival of living organisms. (p. 94)

niche: in an ecosystem, refers to the unique ways an organism survives, obtains food and shelter, and avoids danger. (p. 23)

nitrogen cycle: model describing how nitrogen moves from the atmosphere to the soil, to living organisms, and then back to the atmosphere. (p. 46)

nitrogen fixation: process in which some types of bacteria in the soil change nitrogen gas into a form of nitrogen that plants can use. (p. 46)

nonrenewable resources: natural resources, such as petroleum, minerals, and metals, that are used more quickly than they can be replaced by natural processes. (p. 95)

nuclear energy: energy produced from the splitting apart of billions of uranium nuclei by a nuclear fission reaction. (p. 98)

especies nativas: organismos originales de un ecosistema. (p. 134)

recursos naturales: partes del medio ambiente terrestre que proporcionan materiales útiles o necesarios para la supervivencia de los organismos vivos. (p. 94)

nicho: en un ecosistema, se refiere a las formas únicas en las que un organismo sobrevive, obtiene alimento, refugio y evita el peligro. (p. 23)

ciclo del nitrógeno: modelo que describe cómo se mueve el nitrógeno de la atmósfera al suelo, a los organismos vivos y de nuevo a la atmósfera. (p. 46)

fijación del nitrógeno: proceso en el cual algunos tipos de bacterias en el suelo transforman el nitrógeno gaseoso en una forma de nitrógeno que las plantas pueden usar. (p. 46)

recursos no renovables: recursos naturales, como el petróleo, los minerales y los metales, que son utilizados más rápidamente de lo que pueden ser reemplazados mediante procesos naturales. (p. 95)

energía nuclear: energía producida a partir del fraccionamiento de billones de núcleos de uranio mediante una reacción de fisión nuclear. (p. 98)

O

ozone depletion: thinning of Earth's ozone layer caused by chlorofluorocarbons (CFCs) leaking into the air and reacting chemically with ozone, breaking the ozone molecules apart. (pp. 105, 136)

agotamiento del ozono: reducción de la capa de ozono causada por clorofluorocarbonos (CFCs) que se liberan al aire y reaccionan químicamente con el ozono descomponiendo sus moléculas. (pp. 105, 136)

P

parasitism: a type of symbiotic relationship in which one organism benefits and the other organism is harmed. (p. 22)

petroleum: nonrenewable resource formed over hundreds of millions of years mostly from the remains of microscopic marine organisms buried in Earth's crust. (p. 95)

pioneer species: first organisms to grow in new or disturbed areas. (p. 64)

pollutant: substance that contaminates any part of the environment. (p. 102)

population: all the organisms that belong to the same species living in a community. (p. 10)

parasitismo: tipo de relación simbiótica en la que un organismo se beneficia y el otro es perjudicado. (p. 22)

petróleo: recurso no renovable formado durante cientos de millones de años, en su mayoría a partir de los restos de organismos marinos microscópicos sepultados en la corteza terrestre. (p. 95)

especies pioneras: primeros organismos que crecen en áreas nuevas o alteradas. (p. 64)

contaminante: sustancia que contamina cualquier parte del medio ambiente. (p. 102)

población: todos los organismos que pertenecen a la misma especie dentro de una comunidad. (p. 10)

producer: organism, such as a green plant or alga, that uses an outside source of energy like the Sun to create energy-rich food molecules. (p. 20)

productor: organismo, como una planta o un alga verde, que utiliza una fuente externa de energía, como la luz solar, para producir moléculas de nutrientes ricas en energía. (p. 20)

R

recycling: conservation method that is a form of reuse and requires changing or reprocessing an item or natural resource. (p. 113)

reintroduction program: conservation strategy that returns organisms to an area where the species once lived and may involve seed banks, captive populations, and relocation. (p. 142)

renewable resources: natural resources, such as water, sunlight, and crops, that are constantly being recycled or replaced by nature. (p. 95)

reciclaje: método de conservación como una forma de reutilización y que requiere del cambio o reprocesamiento del producto o recurso natural. (p. 113)

programa de reintroducción: estrategia de conservación que devuelve a los organismos a un área en la que la especie vivió alguna vez, pudiendo involucrar bancos de semillas, poblaciones cautivas y reubicación. (p. 142)

recursos renovables: recursos naturales, como el agua, la luz solar y los cultivos, que son reciclados o reemplazados constantemente por la naturaleza. (p. 95)

S

soil: mixture of mineral and rock particles, the remains of dead organisms, air, and water that forms the topmost layer of Earth's crust and supports plant growth. (p. 38)

succession: natural, gradual changes in the types of species that live in an area; can be primary or secondary. (p. 64)

symbiosis: any close relationship between species, including mutualism, commensalism, and parasitism. (p. 22)

suelo: mezcla de partículas minerales y rocas, restos de organismos muertos, aire y del agua que forma la capa superior de la corteza terrestre y favorece el crecimiento de las plantas. (p. 38)

sucesión: cambios graduales y naturales en los tipos de especies que viven en un área; puede ser primaria o secundaria. (p. 64)

simbiosis: cualquier relación estrecha entre especies, incluyendo mutualismo, comensalismo y parasitismo. (p. 22)

T

taiga (TI guh): world's largest biome, located south of the tundra between 50° N and 60° N latitude; has long, cold winters, precipitation between 35 cm and 100 cm each year, cone-bearing evergreen trees, and dense forests. (p. 70)

temperate deciduous forest: biome usually having four distinct seasons, annual precipitation between 75 cm and 150 cm, and climax communities of deciduous trees. (p. 71)

temperate rain forest: biome with 200 cm to 400 cm of precipitation each year, average temperatures between 9°C and 12°C, and forests dominated by trees with needlelike leaves. (p. 71)

taiga: el bioma más grande del mundo, localizado al sur de la tundra entre 50° y 60° de latitud norte; tiene inviernos prolongados y fríos, una precipitación que alcanza entre 35 y 100 centímetros cúbicos al año, coníferas perennifolias y bosques espesos. (p. 70)

bosque caducifolio templado: bioma que generalmente tiene cuatro estaciones distintas, con una precipitación anual entre 75 y 150 centímetros cúbicos y un clímax comunitario de árboles caducifolios. (p. 71)

bosque lluvioso templado: bioma con 200 a 400 centímetros cúbicos de precipitación al año; tiene una temperatura promedio entre 9 y 12°C y bosques dominados por árboles de hojas aciculares. (p. 71)

threatened species: species that is likely to become endangered in the near future. (p. 131)

tropical rain forest: most biologically diverse biome; has an average temperature of 25°C and receives between 200 cm and 600 cm of precipitation each year. (p. 72)

tundra: cold, dry, treeless biome with less than 25 cm of precipitation each year, a short growing season, permafrost, and winters that can be six to nine months long. Tundra is separated into two types: arctic tundra and alpine tundra. (p. 69)

especies amenazadas: especies susceptibles de verse amenazadas en un futuro cercano. (p. 131)

selva húmeda tropical: el bioma más diverso biológicamente; tiene una temperatura promedio de 25°C y recibe entre 200 y 600 centímetros cúbicos de precipitación al año. (p. 72)

tundra: bioma sin árboles, frío y seco, con menos de 25 centímetros cúbicos de precipitación al año; tiene una estación corta de crecimiento y permafrost e inviernos que pueden durar entre 6 y 9 meses. La tundra se divide en dos tipos: tundra ártica y tundra alpina. (p. 69)

W

water cycle: model describing how water moves from Earth's surface to the atmosphere and back to the surface again through evaporation, condensation, and precipitation. (p. 45)

wetland: a land region that is wet most or all of the year. (p. 79)

ciclo del agua: modelo que describe cómo se mueve el agua de la superficie de la Tierra hacia la atmósfera y nuevamente hacia la superficie terrestre a través de la evaporación, la condensación y la precipitación. (p. 45)

zona húmeda: región lluviosa la mayor parte del año. (p. 79)

> *Italic numbers = illustration/photo*   **Bold numbers = vocabulary term**
> *lab = indicates a page on which the entry is used in a lab*
> *act = indicates a page on which the entry is used in an activity*

Index

Magnification Key: Magnifications listed are the magnifications at which images were originally photographed.
LM–Light Microscope
SEM–Scanning Electron Microscope
TEM–Transmission Electron Microscope

Acknowledgments: Glencoe would like to acknowledge the artists and agencies who participated in illustrating this program: Absolute Science Illustration; Andrew Evansen; Argosy; Articulate Graphics; Craig Attebery represented by Frank & Jeff Lavaty; CHK America; John Edwards and Associates; Gagliano Graphics; Pedro Julio Gonzalez represented by Melissa Turk & The Artist Network; Robert Hynes represented by Mendola Ltd.; Morgan Cain & Associates; JTH Illustration; Laurie O'Keefe; Matthew Pippin represented by Beranbaum Artist's Representative; Precision Graphics; Publisher's Art; Rolin Graphics, Inc.; Wendy Smith represented by Melissa Turk & The Artist Network; Kevin Torline represented by Berendsen and Associates, Inc.; WILDlife ART; Phil Wilson represented by Cliff Knecht Artist Representative; Zoo Botanica.

Photo Credits

Cover Darrell Gulin/Getty Images; **i ii** Darrell Gulin/Getty Images; **iv** (bkgd)John Evans, (inset)Darrell Gulin/Getty Images; **v** (t)PhotoDisc, (b)John Evans; **vi** (l)John Evans, (r)Geoff Butler; **vii** (l)John Evans, (r)PhotoDisc; **viii** PhotoDisc; **ix** Aaron Haupt Photography; **x** Lynn M. Stone/DRK Photo; **xi** Hal Beral/Visuals Unlimited; **xii** (l)Michael P. Gadomski/Photo Researchers, (r)Zig Leszczynski/Earth Scenes; **1** (l)Rob & Ann Simpson/Visuals Unlimited, (r)Steve Wolper/DRK Photo; **2** (t)Darren Bennett/Animals Animals, (b)Collection of Glenbow Museum, Calgary, Canada; **3** (t)Mathew Cavanaugh/AP/Wide World Photos, (b)Helen Hardin 1971; **4** (t)Grant Heilman Photography, Inc., (b)Rick Poley/Visuals Unlimited, Inc.; **5** (t)Jeff Cooper/The Salina Journal/Associated Press, (b)Fletcher & Baylus/Photo Researchers; **6–7** Joe McDonald/Visuals Unlimited; **8** (tr)Richard Kolar/Animals Animals, (l)Adam Jones/Photo Researchers, (c)Tom Van Sant/Geosphere Project, Santa Monica/Science Photo Library/Photo Researchers, (br)G. Carleton Ray/Photo Researchers; **9** (t)John W. Bova/Photo Researchers, (b)David Young/Tom Stack & Assoc.; **11** (l)Zig Leszczynski/Animals Animals, (r)Gary W. Carter/Visuals Unlimited; **14** Mitsuaki Iwago/Minden Pictures; **15** Joel Sartore from Grant Heilman; **17** (t)Norm Thomas/Photo Researchers, (b)Maresa Pryor/Earth Scenes; **18** (tl)Wyman P. Meinzer, (r)Bud Neilson/Words & Pictures/PictureQuest, (bl)Wyman P. Meinzer; **20** (l)Michael Abbey/Photo Researchers, (r)OSF/Animals Animals, (b)Michael P. Gadomski/Photo Researchers; **21** (tl)William J. Weber, (tlc)William J. Weber, (tcr)Lynn M. Stone, (tr)William J. Weber, (bl)Larry Kimball/Visuals Unlimited, (blc)George D. Lepp/Photo Researchers, (bcr)Stephen J. Krasemann/Peter Arnold, Inc., (br)Mark Steinmetz; **22** (t)Milton Rand/Tom Stack & Assoc., (c)Marian Bacon/Animals Animals, (b)Sinclair Stammers/Science Photo Library/Photo Researchers; **23** (tl)Raymond A. Mendez/Animals Animals, (bl)Donald Specker/Animals Animals, (br)Joe McDonald/Animals Animals; **24** Ted Levin/Animals Animals; **25** Richard L. Carlton/Photo Researchers; **26** (t)Jean Claude Revy/PhotoTake, NYC, (b)OSF/Animals Animals;

27 Runk/Schoenberger from Grant Heilman; **28** Eric Larravadieu/Stone/Getty Images; **29** (l)C.K. Lorenz/Photo Researchers, (r)Hans Pfletschinger/Peter Arnold, Inc.; **30** CORBIS; **32** (l)Michael P. Gadomski/Photo Researchers, (r)William J. Weber; **34–35** Ron Thomas/Getty Images; **36** Kenneth Murray/Photo Researchers; **37** (t)Jerry L. Ferrara/Photo Researchers, (b)Art Wolfe/Photo Researchers; **38** (t)Telegraph Colour Library/FPG/Getty Images, (b)Hal Beral/Visuals Unlimited; **39** (l)Fritz Polking/Visuals Unlimited, (r)R. Arndt/Visuals Unlimited; **40** Tom Uhlman/Visuals Unlimited; **44** Jim Grattan; **47** (t)Rob & Ann Simpson/Visuals Unlimited, (c b)Runk/Schoenberger from Grant Heilman; **50** WHOI/Visuals Unlimited; **54** Gerald and Buff Corsi/Visuals Unlimited; **55** Jeff J. Daly/Visuals Unlimited; **56** Gordon Wiltsie/Peter Arnold, Inc.; **57** (l)Soames Summerhay/Photo Researchers, (r)Tom Uhlman/Visuals Unlimited; **62–63** William Campbell/CORBIS Sygma; **64** Jeff Greenberg/Visuals Unlimited; **65** Larry Ulrich/DRK Photo; **66** (bkgd)Craig Fujii/Seattle Times, (l)Kevin R. Morris/CORBIS, (tr br)Jeff Henry; **67** Rod Planck/Photo Researchers; **69** (t)Steve McCutcheon/Visuals Unlimited, (bl)Pat O'Hara/DRK Photo, (br)Erwin & Peggy Bauer/Tom Stack & Assoc.; **70** (tl)Peter Ziminski/Visuals Unlimited, (c)Leonard Rue III/Visuals Unlimited, (bl)C.C. Lockwood/DRK Photo, (br)Larry Ulrich/DRK Photo; **71** (t)Fritz Polking/Visuals Unlimited, (b)William Grenfell/Visuals Unlimited; **72** Lynn M. Stone/DRK Photo; **74** (l)McDonald Wildlife Photography, Inc./DRK Photo, (r)Steve Solum/Bruce Coleman, Inc.; **75** Kevin Schafer; **77** W. Banaszewski/Visuals Unlimited; **78** (l)Dwight Kuhn, (r)Mark E. Gibson/Visuals Unlimited; **79** James R. Fisher/DRK Photo; **80** D. Foster/WHOI/Visuals Unlimited; **81** (l)C.C. Lockwood/Bruce Coleman, Inc., (r)Steve Wolper/DRK Photo; **82** (tl)Dwight Kuhn, (tr)Glenn Oliver/Visuals Unlimited, (b)Stephen J. Krasemann/DRK Photo; **83** (l)John Kaprielian/Photo Researchers, (r)Jerry Sarapochiello/Bruce Coleman, Inc.; **84** (t)Dwight Kuhn, (b)John Gerlach/DRK Photo; **85** Fritz Polking/Bruce Coleman, Inc.; **86** courtesy Albuquerque Public Schools; **87** (l)James P. Rowan/DRK Photo, (r)John Shaw/Tom Stack & Assoc.; **91** (l)Leonard Rue III/Visuals Unlimited, (r)Joe McDonald/DRK Photo; **92–93** Grant Heilman Photography; **94** (l)Keith Lanpher/Liaison Agency/Getty Images, (r)Richard Thatcher/David R. Frazier Photolibrary; **95** (t)Solar Cookers International, (bl)Brian F. Peterson/The Stock Market/CORBIS, (br)Ron Kimball Photography; **96** Larry Mayer/Liaison Agency/Getty Images; **99** (tr)Torleif Svenson/The Stock Market/CORBIS, (bl)Rob Williamson, (br)Les Gibbon/Cordaiy Photo Library Ltd./CORBIS; **100** Sean Justice; **101** (t)Lowell Georgia/Science Source/Photo Researchers, (cl)NASA, (c)CORBIS, (cr)Sean Sprague/Impact Visuals/PictureQuest, (bl)Lee Foster/Bruce Coleman, Inc., (br)Robert Perron; **102** Philippe Renault/Liaison Agency/Getty Images; **103** (l)NYC Parks Photo Archive/Fundamental Photographs, (r)Kristen Brochmann/Fundamental Photographs; **107** (l)Jeremy Walker/Science Photo Library/Photo Researchers, (c)John Colwell from Grant Heilman, (r)Telegraph Colour Library/FPG/Getty Images; **108** Wilford Haven/Liaison Agency/Getty Images; **109** (tl)Larry Mayer/Liaison Agency/Getty Images, (tr)ChromoSohm/The Stock Market/CORBIS, (cr)David R. Frazier Photolibrary, (br)Inga Spence/Visuals Unlimited; **110** (r)Andrew Holbrooke/The Stock Market/CORBIS, (Paint Cans)Amanita Pictures, (Turpantine, Paint thinner, epoxy)Icon Images,